EMERGENCY PREPAREDNESS

for Health Professionals

EMERGENCY PREPAREDNESS

for Health Professionals

Introduction to Disaster Response

Linda Young Landesman, DrPH, MSW

St. Paul • Los Angeles • Indianapolis

Acquisitions Editor Alison Brown Cerier
Development Editor Spencer J. Cotkin
Production Editor Bob Dreas
Cover and Text Designer Leslie Anderson
Photo Researcher Terri Miller / E-Visual Communications, Inc.
Copy Editor Judy Peterson
Desktop Production Leslie Anderson, Charles Sawyer
Proofreader DPS Associates
Indexer DPS Associates

Photo Credits: Cover image by Marvin Nauman/FEMA. All photo credits are included on page 196.

ISBN 978-0-76383-397-8

875 Montreal Way
St. Paul, MN 55102
E-mail: educate@emcp.com
Web site: www.emcp.com

Printed in the United States of America

16 15 14 13 12 11 10 4 5 6 7 8 9 10

Brief Contents

Contents

To the Student

Disasters such as the World Trade Center attacks on September 11, 2001, and Hurricane Katrina in 2005 have made emergency preparedness a national priority. Government agencies and the healthcare system are preparing for future emergencies, including the threat of pandemic influenza. Meanwhile, on a local level, communities are improving their preparations for emergencies that are common in their areas, such as tornadoes and winter storms.

When you become a health professional, what will this mean for you? What will your role be in emergency response? What do you need to know?

Health professionals have very important roles in emergency response. Through this book, you will learn how to prepare for emergencies and how to apply your skills in this new and often challenging context. You will be prepared to meet your professional responsibilities to your employer, your patients, and your community.

What You Will Learn

CHAPTER 1, **Disasters and Healthcare**, explains the impact of natural and man-made disasters on public health and the healthcare system.

CHAPTER 2 discusses **Emergency Management** on the local and regional levels, the national level, and in hospitals and healthcare facilities. This chapter describes who organizes the response and how the response is organized.

CHAPTER 3 recommends steps toward **Personal Preparedness**, including developing skills, creating family emergency plans, and exploring volunteer opportunities.

CHAPTER 4 introduces **Emergency Response Procedures** in hospitals and other facilities or provisional sites. Some of the procedures that are covered include decontaminating patients, evacuating a health facility, investigating suspicious specimens in the lab, and more.

CHAPTER 5 discusses how health professionals can support **Continuity of Care** for patients with chronic health problems before and after a disaster. Patients with chronic health problems are particularly vulnerable in a disaster.

Each allied health profession has special **Skills and Roles** for emergency response. In CHAPTER 6, you will briefly learn about your profession and the roles of other health professionals on a response team.

Textbook Features for the Student

This textbook contains many features to help you learn about emergency preparedness. Consider how you might use each of the following chapter and texbook elements to its fullest advantage:

- A list of learning objectives at the beginning of each chapter will help you focus on what you need to know.
- *Key Terms* and phrases are set in bold and defined in the text or in boxes and are listed at the end of each chapter. They also are defined in the Glossary at the end of the textbook.
- *A Closer Look* boxes add depth or detail with lists, facts, or examples.
- Each chapter ends with a *Summary* and several different types of questions to help you review what you have learned.
- A list of useful *Websites* is included at the back of the book. Because emergency response is a rapidly evolving field, the Internet is an important source of current information. Specialized skills and knowledge for particular health professions also may be found in these websites.

- An *Abbreviations* box at the beginning of each chapter will help you learn the terms, agencies, and procedures that are commonly referred to by their abbreviations. These abbreviations give health professionals and others who work in emergency response a common language. A table of all abbreviations is included at the end of the book.

- The information for this textbook was gathered from a wide variety of sources, including experts in the field, journal articles, newspaper and magazine sources, books, government documents, and websites. Key references and list of resources are provided at the end of the textbook in the *References and Resources* section.

To the Instructor

Emergency Preparedness for Health Professionals has been developed to prepare all allied health students to:

- Play their important roles in emergency response during their careers
- Meet the expectations of their future employers
- Volunteer effectively and serve their communities
- Be confident and safe responders

Students will learn how to apply what they already know and can do in a new and challenging context. The approach focuses on understanding the context of emergency response.

This textbook looks at response from the point of view of the technician or assistant, not the physician or health administrator. Examples of such fields are medical assisting, pharmacy technician, laboratory technician, medical imaging and radiation therapy, surgical technology, physical therapist assistant, health information technician, and EMT. The book can also be adopted for other programs including nursing, licensed practical nursing, and public health.

Many accrediting bodies in allied health are considering the addition of a new standard or set of competency requirements for emergency preparedness. This textbook can be used to meet such requirements.

This textbook can be used as the primary text for a separate course on emergency preparedness for health professionals. Because the material incorporates an all-professions approach, it will work well for a variety of courses in various allied health programs. Emergency preparedness is an ideal choice for creating an interprofessional course, which is a goal of many schools.

This textbook was intentionally kept relatively brief so that it would work well as a practical supplement for a variety of allied health courses. Such courses include:

- Introduction to Health Care
- Healthcare Systems
- Community Health
- Ethics and Law
- Clinical Seminars
- Surgical Technology: Surgical Procedures or Special Topics
- Pharmacy Technician: Pharmacy Practice
- Respiratory Therapy: Emergency and Special Procedures
- Medical Assisting: Administrative Procedures
- Laboratory Technology: Clinical Microbiology
- Health Information Technology: Medical Records
- Emergency Medical Technician: Introduction

The goal of this textbook is to introduce all allied health students to the same body of knowledge on emergency preparedness. This all-professions approach was chosen because students in all fields need the same foundation, and will all need to work together in emergency response. The final chapter briefly introduces the special skills and considerations of individual professions.

The publisher's Internet Resource Center provides instructors with a variety of resources, including course outlines, teaching suggestions, and a reportable test. Contact your sales representative for access to this website.

Acknowledgements

The author and editorial staff at Paradigm Publishing would like to thank the following reviewers for their comments and suggestions:

John H. Clouse, MSR, RT
University of Louisville
Louisville, KY

George Fakhoury, MD, DORCP, CMA
Heald College
San Francisco, CA

Beverly Hawkins, CPhT
Chattanooga State Technical Community College
Chattanooga, TN

D. Randy Kuykendall, MLS, NREMT-P
Colorado Department of Public Health and Environment
Colorado Springs, CO

Michele G. Miller, MEd, CMA
Lakeland Community College
Kirtland, OH

Sandra Namio, CST/CFA
Mohave Community College
Lake Havasu City, AZ

Lara Skaggs
Oklahoma Department of Career and Technology Education
Stillwater, OK

Ruth Thompson, MS, RRT
West Kentucky Community & Technical College
Paducah, KY

Kerry Weinberg, MPA, RT, RDMS, RDCS
New York University
New York, NY

James Willis III
National College
Salem, VA

Grant Wilson, MEd, CST
Calhoun Community College
Decatur, AL

Lori A. Woeste, EdD
Illinois State University
Normal, IL

Lana Zinger, PhD
Queensborough Community College
New York, NY

Furthermore, we would like to thank Roberta Mantus and Judy Peacock for contributing to the presentation and pedagogy of the book.

Author's Note of Appreciation

Because emergency preparedness material has not been previously framed or organized for this audience, researching this book took me to the experts in the field and to those who are incorporating allied health personnel into planning for their own communities. I am indebted to the professionals who shared their work and thank them for their generosity and collegiality.

- My colleagues at the New York City Department of Health and Mental Hygiene provided valuable details. Doug Ball and Anne Rinchiuso generously shared "in the field" descriptions of the Medical Reserve Corps, how volunteers are recruited, and how their skills can be utilized in an emergency. Gary Beaudry imparted an insightful assessment of the role of the laboratory.
- Van Dunn, New York City Health and Hospitals Corporation, contributed a comprehensive description of the tasks of the various health professionals in emergency preparedness. Adrian Dandrea highlighted problems associated with medical records.
- Melissa Corrado and Janet Patry, Primary Care Development Corporation, gave an in-depth understanding of the role of healthcare professionals in ambulatory care settings.
- Linda Degutis sent me in the right direction.

Several colleagues provided specific guidance for their individual professions.

- From Columbia University, Roberta Locko and Stephanie Evans furnished core resources on the role of medical imagers and radiation technologists. Ernestine Pantel described the role of occupational and physical therapists in hospital preparedness.
- Anne Covey of Memorial Sloan Cancer Center recalled the handling of patients in the radiology department during the blackout of 2004.

- Connie Doyle, a DMAT team member and a physician pioneer who championed the involvement of health professionals in emergency preparedness, shared her experiences with the NDMS.

- Scott Becker, Megan Latshaw, and Burt Wilcke of the Association of Public Health Laboratories laid out the foundation for the role of laboratories.

- Grant Wilson, Program Director for Surgical Technology at Calhoun Community College, shared information about the role of surgical technologists.

Special thanks to the Acquisitions Editor at Paradigm Publishing, Alison Brown Cerier, for her vision and understanding of the need for this book.

My ability to deliver the manuscript on schedule was due to my husband Paul taking on more than his share, accepting the piles of resources that took over three rooms, and patiently waiting for the new book to be done. His singular contribution made this book possible.

About the Author

Linda Young Landesman, DrPH, MSW, is a national expert on the role of public health in disaster preparedness and response. She has authored and edited six books, including *Public Health Management of Disasters: The Practice Guide,* and has developed national standards for emergency medical services response. Dr. Landesman was the Principal Investigator for the first curriculum on the public health management of disasters, developed through a cooperative agreement with the Association of Schools of Public Health and sponsored by the Centers for Disease Control and Prevention. This curriculum is currently being used nationwide.

Since 1996, Dr. Landesman has been an assistant vice president at the New York City Health and Hospitals Corporation. She has been appointed to the Weapons of Mass Destruction Committee of the New York City Department of Health and Mental Hygiene, the Advisory Committee of the Mailman School Center for Public Health Preparedness, the Advisory Committee of the Mailman School World Trade Center Evacuation study, and the Emergency Preparedness Council of the New York City Health and Hospitals Corporation. Dr. Landesman was appointed by the New York State Assembly to the Hudson Valley Regional Advisory Committee, Commission on Health Care Facilities in the 21st Century. In addition, she serves on the Commissioner's Community Advisory Committee of the New York City Department of Health and Mental Hygiene. Dr. Landesman is an elected member of the Executive Board of the American Public Health Association and has served on the editorial board of the *American Journal of Public Health.*

Dr. Landesman earned her BA and MSW degrees from the University of Michigan and practiced clinical social work for 10 years. She received her DrPH in health policy and management from the Columbia School of Public Health. Her doctoral dissertation focused on hospital preparedness for chemical accidents and won the Doctoral Dissertation Award from the Health Services

Improvement Fund in 1990. Dr. Landesman has had faculty appointments at the New School, Albert Einstein College of Medicine, and Mailman School of Public Health at Columbia University. She is now on the faculty of the Public Health Practice Program at the University of Massachusetts–Amherst, where she teaches public health emergency management online.

CHAPTER 1

Disasters and Healthcare

What you will learn:

- How health professionals are preparing for their role in emergency response
- Ways that disasters impact public health
- Ways that disasters challenge the healthcare system and health professionals

Imagine that an emergency strikes your community today. You are a health professional working in a hospital, long-term care facility, medical office, or clinic. What should you expect? Will you know what to do?

This book will prepare you to perform your part as a health professional in emergency preparedness and response. You will learn how to respond effectively, safely, and confidently. This chapter will provide a basic orientation to emergency preparedness and will describe the many ways that disasters impact both public health and healthcare.

Disasters can occur on enormous scales. On December 26, 2004, a powerful earthquake occurred under the Indian Ocean that produced a tsunami that devastated large portions of coastal areas of Asia and Africa. In the Aceh province of Indonesia, shown here, more than 100,000 people were killed and approximately 400,000 were left homeless. In the United States, the states that border the Pacific Ocean are the most vulnerable to tsunamis.

An Orientation

What is a disaster? What do employers and communities expect health professionals to know about emergency response? Let's start with the basics.

Basic Terms

Emergencies are incidents that threaten public safety, health, and welfare. Some emergencies are limited, such as a house fire. Others are larger in scope and require a more rigorous response. Examples include snowstorms, tornadoes, and earthquakes. Hospital staff often use the term "emergency" to mean a situation in which a

Abbreviations

CBRN	chemical, biological, radiological, nuclear
CDC	Centers for Disease Control and Prevention
FEMA	Federal Emergency Management Agency
NDMS	National Disaster Medical System
SARS	severe acute respiratory syndrome

patient requires immediate care or a higher level of care. As in the field of public health, this book defines emergency as a broader incident involving groups of people.

An emergency that requires outside assistance is defined as a **disaster**. Responses to emergencies always begin at the local level. Local government agencies manage the response, which may involve the fire department, the police, and the local healthcare system. When an emergency has a serious or widespread impact on an area, the **community**—which here is defined as a jurisdiction such as a town, city, metropolitan area, or county—will need help from state and federal governments. Outside assistance will often include healthcare professionals and medical supplies.

Several other terms are commonly used in discussions of how to respond to emergencies. **Emergency preparedness** is the act of making plans to prevent, respond to, and recover from emergencies. Government agencies, healthcare systems, and even families take part in emergency preparedness. **Emergency response** is a coordinated action to meet the needs of communities affected by an emergency. **Victims** are people who are injured or become ill as a result of an emergency or disaster.

All-Hazards Approach

The emergency planning of a community focuses on local **hazards**, events, or situations with the potential to harm people, property, and the natural environment. For example, some areas are prone to flooding, but seldom experience a blizzard. Tornadoes are common in the Midwest and South, but are rare in California, where earthquakes are a serious concern. Other hazards are due to technological mishaps or man-made causes. For example, a community may have an industrial plant that produces and stores chemicals that could spill or explode.

How can communities prepare for so many different emergencies? Although emergencies vary in size and scope, many similarities exist in the types of challenges that different emergencies present. Because of these similarities, the recommended approach is called **all-hazards**. The all-hazards approach provides general preparation and training that can be applied in a wide range of emergency situations. This approach is more effective and practical than training for every potential type of emergency. The all-hazards approach works because many emergencies have similar impacts on health and the healthcare system.

Roles of Health Professionals

Health professionals perform many important functions in the emergency response of a community. Each profession has special clinical and administrative skills that are invaluable in emergency response. Health professionals have basic medical skills that are needed to care for many people in emergencies. They understand how healthcare systems work, and they are skilled at caring for people who are injured, ill, or worried. The most numerous healthcare workers are the allied health professionals. They are an invaluable part of the healthcare system, particularly in emergencies when every hand is needed.

When health professionals respond to an emergency, they do so within their **scope of practice**, which is the legal and professional definition of their training and responsibilities. Scope of practice is critical for maintaining standards of care. In emergencies, healthcare professionals carry out their usual job responsibilities, but often will work faster, harder, and longer. They will be applying their professional skills in a stressful environment. They will also fill many support roles.

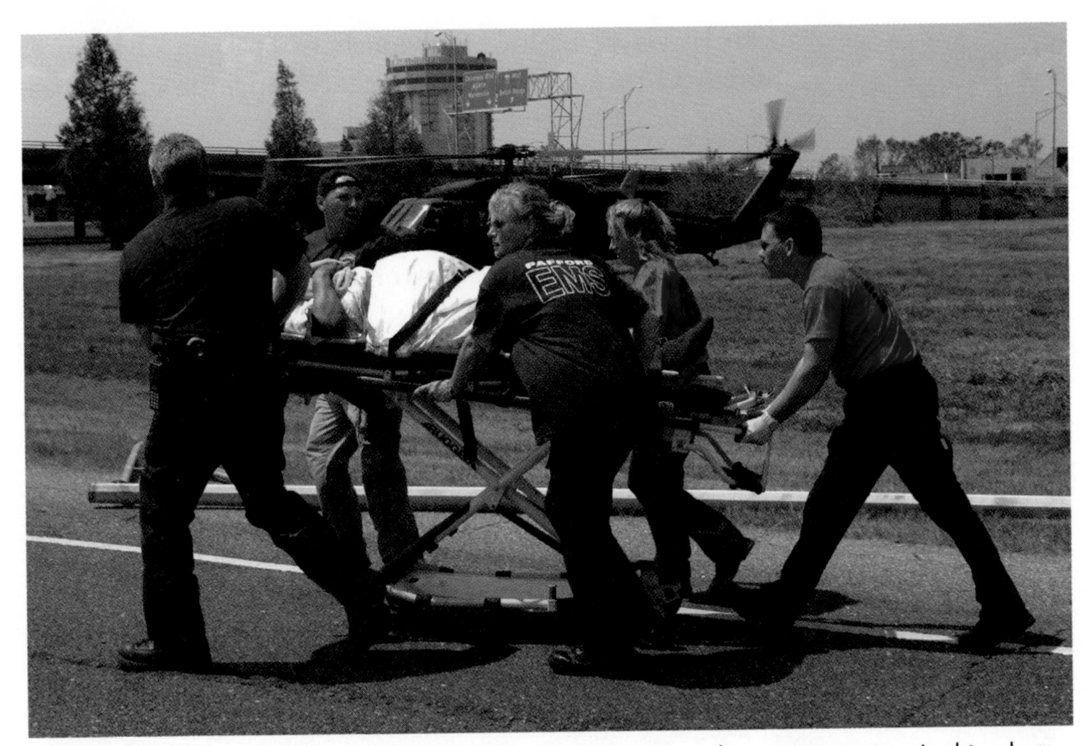

Many different types of professional health workers respond to emergencies. In this photo, emergency workers are moving a rescued evacuee from New Orleans to a medical triage area after he was rescued by a medevac helicopter. After Hurricane Katrina struck the Louisiana and Mississippi coast on August 29, 2005, thousands of people were rescued from flooded areas. Approximately 1,800 people died as a result of the hurricane.

Healthcare professionals are eager to do whatever they can do to help in an emergency. Their concern for people and their professionalism motivate them to help others in need. Participating in emergency response is often stressful, but often deeply fulfilling. It is also being a good medical citizen.

New Expectations

As healthcare systems across the country have come to better understand how to prepare for emergencies, new expectations of their employees have evolved. Hospitals and other health facilities need a workforce that is prepared to fill vital roles in emergency response.

Associations of health professionals have begun to promote training and participation in emergency activities. Educational programs for health professions are also adopting standards and adding course work that will prepare their graduates to meet these new expectations.

By studying and applying the principles discussed in this book, you will build a foundation of knowledge about emergency response. As a healthcare employee, you will find many ways to continue learning about emergency preparedness. Your employers will provide training specific to your workplace. You can learn about emergency preparedness through continuing education offered by professional organizations, schools, and volunteer groups. Additionally, look for resources and specialized courses on the Internet.

Impacts of Natural Disasters

To begin learning about emergency preparedness, think about the many ways that disasters affect the health of a community. Some of these ways are obvious, but others are not. This section explores the impacts of disasters that result from natural causes. The next section examines the impact of disasters that are the result of human actions or activities.

Natural disasters are caused by meteorological or geological events. They include earthquakes, floods, heat waves, hurricanes, tornadoes, winter storms, volcanic eruptions, and wildfires. Disasters vary widely in the severity of damage they cause. Healthcare professionals need to be especially prepared to meet the challenges associated with the disasters most likely to occur in the area in which they live.

Earthquakes

Most earthquakes are produced by movement in the solid, outermost part of the earth, where portions of the earth's crust slide by one another along fractures known as faults. As two parts of the earth rapidly slide by one another, enormous amounts

Earthquakes have affected human civilization since before recorded history. Building collapse is among the most common effects. A powerful earthquake hit central Taiwan on September 23, 1999, destroying many buildings. More than 2,000 people were killed and nearly 8,000 injured. In the United States, thirty-nine states are seismically active, and most people think the Midwest is overdue for a major temblor. During the 19th century, very large earthquakes struck southeastern Missouri and South Carolina.

of energy are released. The energy rapidly travels outward in all directions as waves, similar to how ripples are produced when a pebble is tossed into a pond. The earthquake, or seismic, waves cause violent shaking of the ground. Earthquakes can cause the collapse of buildings and bridges. They can also cause fires when gas lines are broken. Sometimes they set off avalanches, flash floods, and tsunamis (which are also known as tidal waves).

Different factors influence the number of people who are injured or die in earthquakes. Some factors include the proximity of the earthquake to populated areas, the magnitude of the earthquake, the type of housing, and the time of day that the earthquake occurs. People may have multiple injuries due to an earthquake. They may experience lacerations, fractures, and crush injuries during an earthquake, during the collapse of buildings after the earthquake, or during clean-up activity. People pinned under heavy debris can become dehydrated. Many survivors experience high levels of stress, which can affect their health.

Floods

Floods occur more frequently than any other type of disaster. They usually occur when rainfall, extended over several days, causes a river or stream to overflow into the surrounding area. Excessive rainfall may also cause dams or levees to break, releasing huge amounts of water in a short amount of time. People who live in areas that flood regularly often are prepared. However, people who are affected by breaking dams or levees are rarely prepared because these events occur infrequently. Most flood-related deaths occur from **flash flooding**, or flooding that occurs suddenly. People may die when they try to drive, wade, or ride bicycles through water that is flowing very rapidly.

Floods can affect the health of people in the community in many other ways. Sewage systems may overflow into the flood waters. People can develop gastrointestinal and skin disease from exposure to contaminated water. If the electrical power goes out, refrigerated and frozen food and medications will spoil. Then people with chronic health conditions often develop complications because they cannot access their medications or usual medical care.

If evacuations are necessary, infectious diseases can spread through crowded shelters that have inadequate facilities for hygiene.

Heat Waves

When the daytime temperature stays high for several days in a row and does not drop at night, older adults, infants, and people who are obese may have difficulty adjusting to the heat. This can lead to health problems, which are often exacerbated

Major flooding events occur on all continents. During the summer of 1993, much of the upper Midwest was affected by floods along the Mississippi and Missouri Rivers and their tributaries. This photo shows inundated downtown Des Moines, Iowa.

when both the temperature and humidity are high. More people are affected by heat waves in June and July compared to other months, because it is easier to adapt to the heat as summer goes on.

People usually do not get sick at the start of a heat wave. The body adjusts to the heat by varying blood circulation and perspiring. When the body cannot adjust to heat, the body temperature rises. This can lead to illness and death.

Hurricanes

Hurricanes are huge storms with heavy rain and strong winds. They move very quickly in large circular patterns around a quiet center called the eye. Hurricanes can cover a large geographic

Hurricanes pose a serious threat in many parts of the world. In the United States, hurricanes strike the coastlines of the Gulf of Mexico and Atlantic Ocean nearly every year. Strong winds, heavy rains, and flooding destroy buildings and endanger life. In this photo, a yacht club in Pensacola, Florida, shows the fury of Hurricane Ivan, where a 20- to 30-foot high storm surge demolished buildings. Ninety-two people died during Hurricane Ivan in the Caribbean and United States in September 2004.

area, devastating an entire region. Neighboring communities may not be able to help each other. Meteorologists can usually predict, with reasonable accuracy, the timing and location of landfall. Their predictions make it possible to warn communities and advise them to put their disaster plans into effect.

Despite advance warning, residents do not always follow guidelines to prevent injuries and keep their food and water safe. When they do not evacuate or take shelter, people can drown or be injured by electrical wires that have fallen. They can be cut from flying debris or suffer fractures due to falling trees and flying objects. After a hurricane, contaminated water typically causes gastrointestinal, respiratory, or skin diseases. During or

Tornadoes occur across the United States, but are most common in the Midwest and Southeast. Each year the United States averages 800 tornadoes, which kill 80 people and injure approximately 1,500. On May 16, 2007, the city of Greensburg, Kansas, was demolished by an F5 tornado with 200 mph winds. Ten people were killed and sixty were injured.

after a hurricane, people can have heart attacks or other physical problems resulting from stress.

Tornadoes

Tornadoes are one of nature's most severe storms. Tornadoes can develop during a thunderstorm when winds form a funnel shape as they rotate. Wind in some tornadoes has been measured at speeds of more than 300 miles per hour, which is enough force to destroy entire communities. Many communities use a siren system to warn people that a tornado has been sighted, so that they can take shelter. While tornadoes can occur in all the lower 48 states, they occur most often in the Midwest and Southeast from March through May.

Large snowstorms pose many dangers to health. Impassable roads can hinder the ability of rescue crews to reach people. Denver was on accident alert on December 20, 2006, when a blizzard dropped up to 28 inches of snow, slowing the ambulance shown in this photo.

People who have not taken appropriate shelter can be injured from flying debris or from being knocked over or picked up by the high winds. The stress of tornadoes can cause some people to have heart attacks. In the aftermath, people can become ill from unsafe water or food. When electricity is out, people with chronic illness can become more ill. Their medications may go bad or their electrical medical equipment may not work.

Winter Storms

Winter storms bring cold temperatures, ice, snow, and dangerous driving conditions. The National Weather Service can usually predict winter storms. They then provide storm warnings so that communities can prepare.

Transportation accidents are the cause of most deaths during winter storms. People who are exposed to severe weather can suffer frostbite and hypothermia. Carbon monoxide poisoning can occur when people use natural gas, oil, kerosene, wood, or charcoal stoves improperly to heat their homes. People can be injured by snow blowers or have heart attacks when shoveling heavy snow.

Volcanic Eruptions

A volcano erupts when melted rock, called magma, in the interior of the earth is blasted or extruded (like toothpaste) onto the surface of the earth. Gases are commonly mixed with the magma, which produces a buildup of high pressures. When these high pressures can no longer be contained by the solid earth, eruptions occur. Eruptions can cause lava to flow onto the surface of the earth, releasing poisonous gases, or blast enormous quantities of rocks and fine-grained ash into the air. The heat and weight of flowing lava can destroy everything in its path, but lava usually moves slowly enough so that people have time to evacuate safely.

Volcanic ash blasted into the air can affect people hundreds of miles away. The ash can cause respiratory illness and lung

The town of St. Pierre, on the Caribbean island of Martinique, was destroyed by a fast-moving (hundreds of miles per hour) ash flow from the Mt. Pelée volcano. Most of the estimated 31,000 inhabitants of the town were killed within minutes of the eruption. This photo, taken several weeks after the disaster, shows remnants of the town next to the bay. Cloud-enshrouded Mt. Pelée is in the distance. These types of eruptions occur in volcanic provinces worldwide, from Italy to Japan to Indonesia. In the United States, similarly dangerous volcanoes are located in the western United States and Alaska. In 1980, 57 people were killed when Mt. St. Helens in Washington erupted in a manner similar to the famous Mt. Pelée eruption.

On November 13, 1985, a volcanically-induced mudflow buried the town of Armero, Columbia. A 50-foot wall of mud swept down a river valley after the Nevado del Ruiz volcano erupted. More than 23,000 inhabitants were killed and 5,000 were injured. Potential mudflow hazards exist in many parts of the western United States, especially in metropolitan Seattle, which is located down river from Mt. Ranier, one of the largest volcanoes in North America.

damage. It can also contaminate drinking water and interfere with electrical power. People can be burned from the scalding, hot ash. Ash on the ground can be slippery, so people can fall and be injured. Vehicle accidents can occur due to the slippery, foggy driving conditions. Many deaths following a volcanic eruption are caused by suffocation. The most dramatic volcanic events in terms of human losses involve ash blasted laterally from the summit of volcanoes, which can move at hundreds of miles per hour. A related deadly hazard involves volcanically-produced mudflows. These mudflows are caused when the volcanic heat melts snow at the summit of the volcano. The hot water mixes with ash from previous eruptions, producing fast-moving, dangerous rivers of

In August 2007, many parts of southern Greece experienced wildfires that resulted in more than 60 deaths. Many were caught on a highway when their cars were overtaken by a fire. Wildfires are a hazard in many parts of the United States during dry times of the year.

mud. The residents of the city of Seattle are well aware of the hazards of mudflows from nearby Mt. Ranier.

Wildfires

Wildfires, fires in forests and other natural areas, often spread quickly because of windy, dry weather conditions. Once groundcover has been burned away, little is left to hold soil in place on steep slopes and hillsides. Heavy rains following a fire can cause landslides and floods.

Wildfires can cause burns and inhalation injuries, exacerbate respiratory conditions, and trigger stress-related cardiovascular events while people are fighting or fleeing the fire.

Impacts of Man-Made Disasters

Man-made disasters are disasters caused by people or technology. They occur as a result of system failure, accident or negligence, or intentionally as in the case of terrorism. Examples of man-made disasters include chemical spills, explosions, and nuclear accidents. Sometimes a natural disaster, such as a flood, becomes a man-made disaster when flood waters come in contact with toxic chemicals, and then spread them into a community.

Transportation accidents, such as train or airplane accidents, are made-made disasters. Although infrequent, they may result in enough fatalities or serious injuries to overwhelm the available

The collapse of the Interstate 35W bridge over the Mississippi River, near downtown Minneapolis, occurred on August 1, 2007. Thirteen people died and approximately one hundred were injured. In the immediate aftermath of the collapse, the rapid organization of the incident command center and deployment of emergency personnel showed that emergency preparedness was well implemented.

This photo shows the remains of reactor number 4 of the Chernobyl nuclear power plant in the Ukraine (then part of the Soviet Union). On April 26, 1986, a steam explosion caused a nuclear meltdown in the reactor, additional explosions, and a fire. The release of large quantities of radioactive emissions floated over much of central, northern, and eastern Europe. As a direct result of the accident, 56 people died and an estimated 4,000 additional deaths will be caused by cancer.

resources. These accidents or events are known as **mass casualty incidents**. Severe injuries include fractures, burns, lacerations, and crush injuries. The most common injuries are eye injuries, sprains, strains, minor wounds, and eardrum damage.

The United States is experiencing an aging infrastructure that could result in more emergencies similar to the steam pipe explosion during rush-hour in New York City and the freeway bridge collapse in Minneapolis, both in 2007.

Harmful Chemicals

Accidents or acts of terrorism that result in the release of dangerous chemicals can pose a threat to human health and the environment. Chemicals can be corrosive, flammable, or toxic. They can leak from industrial plants or transportation accidents and cause explosions.

Different hazardous chemicals have different effects on health. Typically, chemicals can cause serious respiratory problems, skin diseases, and burns. Some exposures can cause death.

Radiation

Radiation can leak or explode from a nuclear power plant, industrial factory, or nuclear weapon.

Symptoms related to radiation exposure depend on whether the exposure is external or internal. External exposure can result in burns. Internal exposure can affect the cell structure of the body and could result in cancer. The extent of exposure to radioactive particles determines the amount of damage. Small doses of radiation can impact the central nervous system. Large radiation doses can lead to radiation sickness, resulting in multi-system problems involving the peripheral nerves, lungs, and immune system. Decreased protection against bacterial and virus infections can then occur.

Terrorism

The U.S. Department of Defense defines **terrorism** as the unlawful use of, or threatened use of, force or violence against individuals or property to coerce or intimidate governments or societies, often to achieve political, religious, or ideological objectives. Terrorism can take many forms, including the use of chemical, biological, radiological, or nuclear agents. Emergency planners refer to acts of terrorism by the abbreviation CBRN—chemical, biological, radiological, nuclear.

Bioterrorism is the deliberate release of viruses, bacteria, or agents such as Sarin, that can cause illness or death in people, animals, or plants. The viruses or bacteria are typically altered to more likely cause disease, become resistant to current medicines, or readily spread into the environment. Biological agents can be spread through the air, through water, or in food. Some bioterrorism agents, such as the smallpox virus, can be spread from person to person. Other bioterrorism agents, such as anthrax, cannot be spread by this method.

Explosions, such as those caused by bombings, are another kind of terrorist act. Explosives can produce numerous casualties in seconds. People who can walk away from the accident scene can seek treatment at the closest hospital. These hospitals can expect a large number of victims soon after a terrorist strike. For example, in 2004, commuter trains in Madrid were bombed by terrorists. After the bombings, the closest hospital received 272 patients in less than three hours.

Explosions can cause blast lung injury, which causes continual health problems or death. Explosions can injure all parts of the body and can cause burns. Emergency responders can be injured at the site while searching for survivors.

Emerging Infectious Disease

Infectious diseases may accompany disasters. Contagious diseases can be spread by people being displaced from their homes, staying in crowded shelters, and increases in population density. They can also be spread by the disruption of public utilities, the interruption of basic public health services, and compromises to sanitation and hygiene.

Infectious diseases can also themselves become disasters. The sudden worldwide spread of severe acute respiratory syndrome (SARS) brought new attention to the threat of **emerging infectious disease**, a new or previously limited infection that spreads rapidly in the population. West Nile virus is another example of an emerging infectious disease.

A **pandemic** is an infection that spreads rapidly around the world. Recent concerns about a new strain of influenza have led to worldwide preparations for an **influenza pandemic**. Influenza pandemics occur approximately three times per century. The last influenza pandemic occurred in 1968.

A pandemic infection can cause many illnesses and deaths and may quickly overwhelm healthcare systems. Pandemic infections are dangerous because the population has no immunity and their infection processes may not be immediately understood. In a pandemic, many healthcare workers are likely to become ill, making it difficult to provide healthcare for the population.

Influenza can cause secondary complications such as pneumonia, dehydration, and worsening of chronic respiratory and cardiac problems. Public health authorities predict that the next influenza pandemic could infect 30 percent of the U.S. population.

The table below summarizes many types of disasters discussed in this chapter and the possible impact on the health of a community.

Impacts of Disasters

Disaster	Injuries	Medication Loss	Unsafe Food and Water	Respiratory Illness	Skin Diseases	Psychological Related Stress	Gastrointestinal Illness
Chemical	X			X	X	X	
Earthquake	X	X	X			X	
Flood	X	X	X	X	X	X	X
Heat Wave		X		X		X	
Hurricane	X	X	X	X	X	X	X
Pandemic Influenza				X		X	
Radiological			X	X	X	X	X
Terrorism	X			X	X	X	
Tornado	X	X	X			X	
Volcanic Eruption	X	X	X	X	X	X	
Wildfire	X	X	X	X			
Winter Storm	X			X		X	

Disaster Medical Assistance Team (DMAT) tents were set up immediately in front of a hospital in Punta Gorda, Florida, after Hurricane Charley struck on August 13, 2004.

Challenges for the Healthcare System

Health professionals and the entire healthcare system face many challenges when responding to emergencies.

Initiating Response

In an emergency, the first challenge is knowing what has occurred and what response is needed.

- Sometimes the first sign a healthcare facility receives that something has happened is when victims show up for treatment.

- The local authorities may not have or be able to provide complete information to the healthcare facility.
- Despite spotty or incomplete information, some emergencies, such as explosions, require a rapid response from the facility.
- Slowly developing health emergencies, such as an outbreak of an infectious disease, may not be recognized at the outset.
- Facilities may not be able to get help from local authorities, especially during the early stages of an emergency.

Damaged Facilities

A hospital, clinic, long-term care facility, or physician's office may be damaged in an emergency. This damage can happen when an increase in general or specialized services is urgently needed. The medical facility may also be cut off from electricity, water supply, waste disposal, and communication. A modern healthcare facility greatly depends on these services. Some of the damages to facilities that can occur are:

- The building may be damaged. It may continue to operate, or it may need to be evacuated or closed.
- Power may be lost. Radiology machines, various monitoring devices, life support equipment, and sterilization equipment all require electricity.
- Many items in a healthcare facility are hazardous if they are overturned or damaged, including drugs, chemicals, heavy equipment, and radiation devices.
- Telephone or Internet communications may be disrupted, which will affect both internal and external communications, such as reporting lab results.
- In an emergency that lasts a long time, healthcare facilities may become isolated. Essential supplies such as pharmaceuticals, splints, and bandages may run out without outside assistance.

- The facility may not have enough food and safe water for everyone who is housed there.

Patient Surge

Depending on the kind of emergency that takes place, many people may need medical attention simultaneously. A sudden increase in the number of patients is known as a **patient surge**. The following events can occur in a patient surge:

- Victims may begin arriving on their own soon after the emergency. The facility may not be prepared to receive or treat them.
- Victims will go to the nearest facility, even if it cannot handle the number of people who need medical attention.
- Victims can be all ages, speak different languages, and not be from the area where the emergency occurs.
- Testing and lab facilities may be overwhelmed.
- Some people may not have been physically injured or may not be ill. However, they will come to the facility because they are concerned about what has happened or what might happen to them. These people require evaluation, observation, and perhaps care.
- Patients already in the facility may become very concerned if the emergency involves a contagious disease or a type of contamination.
- Family members, community responders, and the media will come to the facility and will need to be accommodated.

Staff Shortage

During an emergency, medical facilities may need more healthcare professionals than are available. Some staffing issues that may occur during an emergency include:

- Staff may have trouble getting to work.

- Staff may have to work longer hours than normal or work a different schedule.
- Certain kinds of health professionals may be in short supply to meet the increased demands. For example, a medical facility may need more health professionals to handle increased lab work or X-rays and other imaging.
- Staff may not be working with their usual coworkers.
- Professionals from other facilities may be given privileges to help. They may not be familiar with the procedures or facility.

Continuing Health Services

In addition to caring for victims of an emergency, health professionals must continue to provide care to other patients. Elective procedures will be cancelled if disaster strikes, but unrelated acute health problems may occur. The following are some of the unrelated acute health problems that can occur.

- Women will come to the facility to give birth.
- People will have asthma attacks or heart attacks.
- People with chronic illness or special medical needs that require monitoring and medical supervision will need regular care.

Loss of Medical Records

Disasters may damage records at both the healthcare and storage facilities, making them unavailable when they are most needed. Patient records are critical for appropriate patient treatment. Continuity of healthcare is threatened when patients are evacuated to other facilities without a written record of their condition, treatment, and medication and dosages. Paper records may be unavailable or destroyed. Electronic medical records may not be available if power is out or computers are damaged.

Supply Shortages

Healthcare facilities generally maintain a stock of pharmaceuticals, medical supplies, and equipment for daily operations. They generally order enough for normal conditions. In an emergency, healthcare facilities can run out of equipment, equipment parts, and essential medications when many more patients require care.

Personal Situations

Healthcare professionals can also be victims of a disaster. Some of the disaster effects that can occur to healthcare professionals include:

- Their homes may be damaged or destroyed.
- Their families or friends may be injured, or even killed.
- They may become ill or injured.
- They may suffer from stress both during and after the emergency.

Lessons from Hurricane Katrina

The story of Hurricane Katrina shows the many ways that a disaster can affect a city and the many challenges for those responding. Let us look at the impact of this disaster on the health of the citizens of New Orleans and on the viability of the healthcare system of the city. Katrina has many lessons to teach.

Disaster Strikes

In the summer of 2005, sixteen city hospitals were providing inpatient and outpatient care to the people of New Orleans and the surrounding area. New Orleans was the center for the public health system of the state. Charity Hospital was the primary source of healthcare for a largely poor and uninsured population, with

A rescue boat is picking up victims in a flooded neighborhood in New Orleans several days after Hurricane Katrina struck Louisiana in August 2005.

high rates of chronic diseases such as heart disease and diabetes. Charity Hospital had one of the busiest emergency departments in the state.

Hurricane Katrina struck the Gulf Coast on Monday, August 29, 2005. After the hurricane passed, many residents thought the emergency was over. However, New Orleans, located about six feet below sea level, is surrounded by water and is protected by levees. The levees soon broke, and flooding occurred in 80 percent of New Orleans. Approximately 1,800 people lost their lives during Katrina and its aftermath. After the hurricane, an estimated 3 million people were without electricity. Within a month, every state was involved in providing shelter and healthcare to the more than 1 million evacuees.

Impact on Healthcare

Hurricane Katrina and the flooding that followed devastated the healthcare system in New Orleans and left many people without access to medical care. The loss of medical care was unlike anything that had ever been experienced in the United States in modern times. All but three hospitals had experienced severe damage or were completely destroyed. Physical access to the remaining facilities was almost impossible because of flooding.

Facilities that did not evacuate before the storm had extreme difficulty maintaining standards of care. In hospitals that lost power, pulmonary ventilator systems and other medical equipment requiring electricity did not work. Most hospitals lacked an adequate supply of fresh water, safe food, and prescriptions. These shortages lasted for days.

Compounding the problems were staff shortages. Many healthcare workers were being evacuated or forced to leave the area because their homes had been destroyed. In the months following Katrina, not enough healthcare professionals were available to attend to all the needs of patients. Many of the available healthcare professionals worked 18-hour days.

Other problems related to healthcare developed soon. Many people lost their health insurance when employers did not reopen their businesses after the hurricane. People had difficulty finding pharmacies, locating their old doctors or connecting with new ones, and paying for medical care. Medical specialists were difficult to find.

Health System Response

Healthcare professionals faced numerous challenges in caring for Katrina's victims. Federal, state, and local governments; businesses; corporations; the religious community; and other volunteers pitched in to speed relief to Katrina's victims. However, it was a major challenge to keep all the efforts coordinated.

Soon after the hurricane, a temporary emergency department and medical clinic opened at the New Orleans Convention Center. Four mobile clinics were set up. Attempts were made to open clinics in places such as schools. Healthcare professionals had difficulty getting to these clinics because physical access was blocked in some areas, and civil disorder was a problem in others.

During the aftermath of Hurricane Katrina, Methodist Hospital had 600 in-patients. Everything had to be done without power, communications, or X-rays. Supplies were few. Without power, patients requiring ventilators were kept alive by staff members using hand pumps in one-hour shifts. Staff had to carry patients up and down stairs. The temperature was 110° Fahrenheit with high humidity. Workers were dehydrated. The hospital had to feed over 500 "refugee" non-patients in addition to the staff, but there was not enough food and water.

Communication and Coordination

Communication was the greatest obstacle facing the medical community. Working equipment and standardization of communication equipment and response protocols were lacking. The infrastructure for landline phones was destroyed. The majority of cell phones were inoperable because they could not be recharged. Satellite communications were unreliable.

Without a method of communication, the functioning hospitals did not know whether more patients would be arriving or whether they should send medical teams to other areas. Not being able to communicate outside of the hospital threatened the safety of medical staff and the lives of patients. Incapacitated and without supplies, many healthcare professionals struggled to provide care and keep patients alive until help arrived.

Evacuation

Hospitals and other medical facilities are required to have evacuation plans for emergencies. When Hurricane Katrina was approaching landfall, hospital and nursing home administrators had to decide whether or not to evacuate their patients. Evacuations of medical facilities are risky, difficult, and expensive. An evacuation can take 36 to 72 hours and many staff to move

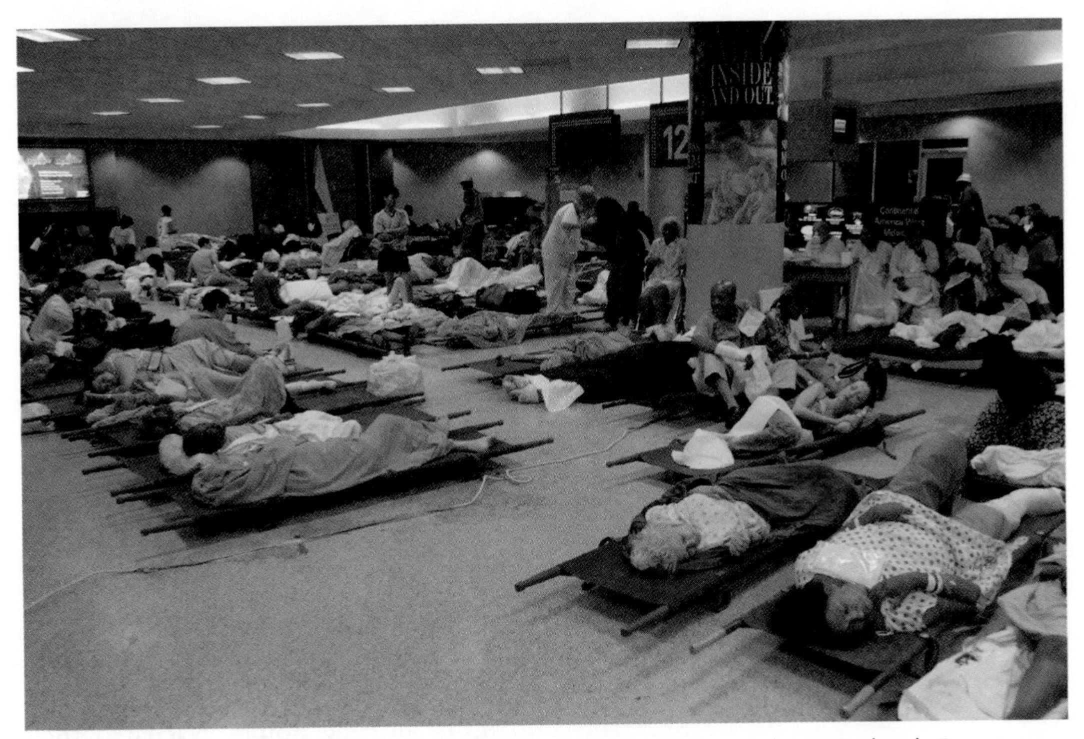

Evacuees and patients arrive at the New Orleans airport where Federal Emergency Management Agency's (FEMA) Disaster Medical Assistance Teams set up operations immediately after Hurricane Katrina struck in August 2005.

patients safely. Many healthcare facilities did not anticipate the flooding. The administrators thought that they were prepared to handle the hurricane without evacuating. In New Orleans, an estimated 215 people died in hospitals and nursing homes as a result of the failure to evacuate.

During the large-scale evacuations from New Orleans that began on September 1, victims from shelters and from failing healthcare facilities were evacuated to a temporary field hospital at the New Orleans airport. These medical response teams were initially overwhelmed, as they evaluated 4,000 evacuees to determine who needed medical attention. After treating the evacuees, teams prioritized and prepared victims for airlift to healthcare facilities outside the flood zone. In addition, over 21,000 displaced persons not needing medical attention were evacuated.

When the hurricane was approaching, hospitals had discharged many patients, but the sickest remained. The medical centers of New Orleans needed to be evacuated after the flood. In previous evacuations such as the one for Hurricane Ivan in 2004, some facilities had moved patients using 18-wheel flatbed trailers equipped with generators and air-conditioning. However, after Hurricane Katrina, facilities that relied on bus companies and ambulance services were unable to access these vehicles because the roads were flooded. Additionally, fuel and rental cars were in short supply, and many forms of public transportation had been shut down before the hurricane arrived.

Because there were no functioning elevators, when the evacuation began, patients had to be carried downstairs, together with their monitoring equipment and IV bags. They were then put into boats when available, and transferred to an ambulance, which took them to the New Orleans Airport. Some patients were taken by helicopter to Florida and Texas. Hospitals evacuated a total of 12,000 people.

Before Katrina's landfall, 19 nursing homes evacuated their residents. Once the flooding began, an additional 32 homes did the same. Administrators at Saint Rita's nursing home, decided to wait out the storm. As a result, 35 people died.

Surge of Patients

Nearby hospitals began to receive the displaced patients. Our Lady of the Lake, the largest hospital in Louisiana, operated at full capacity and hospitalized over 80 displaced patients. All staff members at the hospital worked overtime, often not going home. Some staff filled in for coworkers who could not get to work. The hospital set up a 24-hour child-care center. Medical and other supplies did not arrive until several days after the storm.

Before Katrina, the emergency department of nearby West Jefferson Hospital saw an average of 180 patients a day. Immediately after the storm, the hospital saw as many as 600 patients a day, mostly for nonemergencies. The main problems were wound infections, gastrointestinal problems, dysentery, and dehydration due to lack of clean drinking water. There were many concerns about exposure to toxins and chemicals that mixed with floodwaters. The number of patients hospitalized jumped by 40 percent overnight.

Because West Jefferson Hospital was short of staff, it asked for federal help immediately. A federal team of healthcare professionals arrived one week later. This team immediately erected medical tents outside the emergency department and saw thousands of patients. The team stayed more than two months to help provide needed healthcare to those affected by Katrina.

Equipment and Supplies

Immediately following the hurricane, there was a heavy demand for supplies, although few demands could be met. Lengthy delays and shortages occurred because officials assumed that people would be in hospitals only a short time. Medical equipment and supplies to meet the needs of a disaster of this size had not been delivered to the area in advance.

One of the main distributors of hospital supplies, Cardinal Health, was damaged and closed its distribution center. As a result, supplies that were usually provided locally had to be brought in from Alabama, Georgia, and Texas. Because of security concerns, medical products had to be escorted by local and state authorities.

Once hospitals were able to procure supplies, they ordered two to three times the usual amount because they expected to care for many more patients. Typical requests included water; sand; generators; emergency items such as flashlights, linens and cots; and medical supplies including antibiotics, antibiotic soap, gowns, drapes, and surgical instruments.

Chronic Disease

After the hurricane, many people with chronic illnesses experienced complications because they had no access to essential medications and treatments such as oxygen, insulin, or kidney dialysis.

New Orleans was unprepared to evacuate and provide medical care for **patients with special needs**. These people did not require hospitalization but their physical condition required care that could not be appropriately provided in a general shelter. To complicate the problem, the various agencies and departments that had responsibility for the care of these 45,000 patients could not agree on which patients had special needs.

Medical Records

Hurricane Katrina destroyed millions of pages of patient medical records in medical offices, clinics, and hospitals. Thousands of patients whose medical records were lost had to depend on their memory and knowledge of their medical history when they sought care.

Hospitals such as the Veterans Administration Medical Centers and Kindred Hospital in New Orleans had electronic medical records. These hospitals sent the medical records electronically to facilities with transferred patients.

The lack of medical records contributed to difficulties and delays in the medical treatment of victims. When the Harris County hospital created a large clinic at the Astrodome in Houston, Texas, it included 80 computer terminals for registering patients and recording their medical histories and information. Volunteers in Houston documented patient information and registered 8,000 evacuees, creating new electronic medical records.

Federal Response

The federal government brought in healthcare workers, emergency medical shelters, portable medical units, pharmaceuticals, and supplies.

The U.S. Department of Health and Human Services activated the National Disaster Medical System (NDMS). Patients were transferred out of the flooded area to hospitals that had previously agreed to accept patients in the event of a disaster. NDMS also sent medical teams to provide care and to supplement the few hospitals that were still functioning. These self-sufficient teams brought enough supplies to last for three days. Team members became fatigued because they operated with limited medical supplies, inadequate food and water, intermittent electricity, and no air-conditioning.

The mobile medical facilities were designed to "hospitalize" patients for 24 hours. Following Katrina, these facilities did not keep patients, enabling them to care for 2,500 people in two days.

The Centers for Disease Control and Prevention (CDC) provided millions of doses of vaccines for influenza, tetanus and diphtheria, hepatitis A, and hepatitis B. People handling the storage and distribution of the vaccines had difficulty keeping them refrigerated at the correct temperature, which is essential for safety and effectiveness.

Credentialed volunteers supplemented the local medical services and were an important part of the medical response. The U.S. Public Health Service first looked to the medical volunteers who had preregistered to work in a disaster. The Medical Reserve Corps linked its database of healthcare professionals with state databases to confirm the credentials of the volunteers. Volunteers were urged to work with preexisting rescue teams and not act independently.

Continuing Challenges

Two years after the disaster, seven hospitals in New Orleans have remained closed, including Charity Hospital. The overall number of hospital beds in New Orleans has dropped by more than 50 percent compared to the time just before Hurricane Katrina. The number of physicians, nurses, and other healthcare workers also has declined significantly.

Because of the decline in available healthcare services in the New Orleans area, people with chronic diseases, such as diabetes, hypertension, and respiratory diseases, have continued to struggle. It has been estimated that their death rate has increased by 47 percent compared to patients with the same diseases two years before Hurricane Katrina struck the Gulf Coast.

Summary

Emergency preparedness today takes an all-hazards approach that can be applied in a variety of disasters. Natural disasters are caused by acts of nature. Man-made disasters are caused by people or their technology. Disasters have many expected and unexpected impacts on the health of people in a community. During and after disasters, healthcare facilities may be overwhelmed with victims seeking aid. Facilities may be damaged or unable to operate. Facilities will often experience shortages of supplies if the aftermath of disasters is long-lived. Additionally, facilities may experience staff shortages and loss of medical records. In 2005, Hurricane Katrina and the floods that followed the storm taxed the healthcare system of New Orleans in ways that had never before been experienced in the United States in modern times. Two years later, the city's health facilities have not recovered, and patients are still feeling the impact of the storm's effects.

Key Terms

- all-hazards approach
- bioterrorism
- community
- disaster
- emergencies
- emergency preparedness
- emergency response
- emerging infectious disease
- flash flooding
- hazards
- man-made disasters
- mass casualty incidents
- natural disasters
- pandemic
- patient surge
- patients with special needs
- scope of practice
- terrorism
- victims

Reinforcing Terminology

Choose the word or phrase from the list of key terms that best matches each of the following descriptions.

1. A phenomenon experienced by healthcare facilities following a disaster ________
2. Risks or dangers to humans, property, and the environment ________
3. A type of man-made disaster intended to make people ill ________
4. A mid-air collision of 747s, which results in enough injuries to overwhelm a system, for example ________
5. A new and contagious illness ________
6. Elderly people who die as a result of a prolonged heat wave, for example ________
7. A possible impact of heavy rainfall where rivers overflow ________
8. Unlawful use of force or violence to intimidate or coerce others ________
9. Firefighters, police officers, and healthcare professionals working together to assist victims of a disaster ________
10. An emergency resulting from system failure, accident or negligence, or terror ________

Reviewing Concepts

Choose the response that best answers the question or completes the sentence.

1. A recommended principle for emergency preparedness is
   a. to encourage responders to act outside their scope of practice.
   b. to train responders for specific emergencies.
   c. to recruit responders, or "medical citizens," from the chronically ill.
   d. to prepare responders for a range of emergencies.

2. Which disaster occurs more frequently than any other type of disaster?
   a. tornado
   b. flood
   c. hurricane
   d. earthquake

3. One of the most potentially dangerous results of a volcanic eruption in terms of human loss of life is
   a. an earthquake.
   b. an increase in global warming.
   c. a mudflow.
   d. interruption of ocean shipping.

4. A disaster that could be natural or man-made is
   a. an earthquake.
   b. an oil spill.
   c. a wildfire.
   d. a heat wave.

5. A common health impact of most disasters is
   a. dehydration.
   b. emotional stress.
   c. lacerations.
   d. central nervous system damage.

6. Which disaster is most likely to result in radiation exposure?
   a. nuclear power plant explosion
   b. train accident
   c. infrastructure collapse
   d. fire in a chemical-manufacturing plant

7. Which abbreviation refers to acts of terrorism?
   a. CBRN
   b. NDMS
   c. SARS
   d. FEMA

8. Which factor is likely to make an influenza pandemic more likely?
   a. Medical records are lost or destroyed.
   b. People have no immunity.
   c. Healthcare facilities lose electrical power.
   d. Medical supplies and medications run low.

9. What is often the first challenge for healthcare workers in an emergency?
   a. analyzing why the emergency occurred
   b. seeking assistance from the federal government
   c. assessing the cost of damages
   d. determining what has happened

10. In the aftermath of Hurricane Katrina, what was a major site for federal medical response teams?
    a. Methodist Hospital
    b. Charity Hospital
    c. New Orleans airport
    d. New Orleans convention center

Exploring Emergency Preparedness Issues

1. Go to *www.noaa.gov/wx.html*, the webpage for weather watch and alerts of the National Oceanic and Atmospheric Administration. Click on several of the links, such as Active Weather Alerts. Then click on several of the maps to learn about the weather in your area and around the country. In one or two paragraphs, describe what types of events occur in your community.

2. Go to *www.cdc.gov/mmwr/*, the CDC's Morbidity and Mortality Weekly Report. Type "disasters" in the box next to MMWR Search. Find one case report about a disaster that occurred in your region of the country. Describe the circumstances of that disaster and its impact on the community. What were the health impacts? Describe how your profession would be involved in caring for victims. Summarize this information in a one- to two-page report.

CHAPTER 2

Emergency Management

What you will learn:

- The four phases of comprehensive emergency management: mitigation, preparedness, response, and recovery
- The leading role of the Local Emergency Management Agency
- How Incident Command Systems clarify roles and authority at a disaster site
- How federal disaster aid is initiated and organized; federal resources including the National Disaster Medical System
- What hospitals and other healthcare facilities cover in their emergency operations plans

Who is in charge of planning for emergencies? When a disaster strikes, who organizes and directs the response? A local governmental agency leads the emergency planning and response activities in each community, following guidelines from the federal government. A community is a jurisdiction. It could be a town, city, metropolitan area, or county. All communities and responders follow the same emergency management systems and principles. By following the same management systems, a diverse group of responding organizations, institutions, and government agencies can work together effectively. Understanding how these management systems work will help you see the big picture and your place in it. When you respond to an emergency as an employee or a volunteer, you will be part of an emergency management system and plan.

Comprehensive Emergency Management

Federal, state, and local government agencies, hospitals, private sector companies, and other institutions engage in **comprehensive emergency management**, which means they are involved in four phases of emergency activities: mitigation, preparedness, response, and recovery.

FEMA and New York State Disaster Field Office personnel met in New York City to coordinate federal, state, and local disaster assistance programs soon after the attacks on the World Trade Center on September 11, 2001.

Abbreviations

DMAT	Disaster Medical Assistance Team
DMORT	Disaster Mortuary Operations Response Team
EMS	Emergency Medical Service
EMT	emergency medical technician
EOC	emergency operations center
EOP	emergency operations plan
FEMA	Federal Emergency Management Agency
HICS	Hospital Incident Command System
ICS	Incident Command System
LEMA	Local Emergency Management Agency
NDMS	National Disaster Medical System
NIMS	National Incident Management System
NRF	National Response Framework

Mitigation

Once communities have identified the hazards that could harm people, property, or the environment, they try to prevent or minimize emergencies that could occur from those hazards. This is called **mitigation**. For example, a Florida community might order an evacuation if it learned that a hurricane was heading its way.

Planners also work on reducing the risks posed by the hazard. For example, if an industrial plant produces highly toxic chemicals, a community can pass zoning laws prohibiting the building of residences near the plant. A community might invest in an updated warning system for tornadoes and other weather hazards.

Preparedness

In the **preparedness** phase, planners take actions to enable them to respond to future emergencies effectively. Preparedness includes:

- Organizing personnel, equipment, and supplies so that they can be accessed quickly.
- Training personnel.
- Conducting drills to test the plan.
- Developing communications and warning systems.
- Creating evacuation plans.
- Stockpiling supplies, including medications and medical supplies.
- Arranging agreements with neighboring communities or institutions for mutual assistance, known as **mutual aid**.

Preparedness includes writing an **emergency operations plan (EOP)** that includes procedures and protocols for responding to emergencies. As part of their EOP, communities and institutions plan to establish an **emergency operations center (EOC)** where leaders can gather to coordinate the response.

Response

The **response** phase occurs immediately before, during, or after an emergency event. The response can include fire fighters, police, volunteer organizations, the healthcare system, search and rescue, and others. Response includes search and rescue operations as well as emergency medical care, food, water, and shelter.

Recovery

The **recovery** phase includes activities and programs that help a community return to normal or to what may be called a "new normal." Rebuilding destroyed homes and medical facilities are examples of recovery.

Local personnel are the first to respond to emergencies. Local firefighters are shown in the wreckage of the World Trade Center in September 2001.

Local Response

Local governments have the primary responsibility for emergency management and response. The federal government provides guidelines that communities must follow, but it provides assistance in only one percent of emergencies.

Local Emergency Management Agencies

The **Local Emergency Management Agency** or **Local Emergency Management Authority (LEMA)**, is the community governmental agency with expertise in public safety, emergency medical services, and emergency management. In all emergency management activities, the LEMA has the lead responsibility for coordinating the police, fire department, emergency medical services, public works, public health and healthcare organizations, and utilities.

During an emergency, the LEMA reports to the mayor or city manager. The number of agencies involved and the degree of preparations for an emergency vary by the size of the community and the resources available. In a small town, the chief of police or the local public safety official may be the head of the LEMA. This person works with volunteer emergency medical services and local hospitals. Large metropolitan areas set up an Office of Emergency Management, which works with numerous private and public organizations to ensure that the community is prepared.

The Local Emergency Management Agency develops an emergency operations plan that designates who will be in charge and who will coordinate the public agencies, private sector, and nongovernmental organizations, including healthcare professionals. The plan also identifies what must be done so that communities are prepared for future disasters. This might include such activities as taking inventories of equipment and supplies or training staff.

Many communities establish and maintain an emergency operations center (EOC). When a community executive determines that a disaster is imminent, he or she will activate the EOC of the jurisdiction. Then representatives from all the agencies involved will gather at the center to coordinate the community response.

Local government agencies develop an all-hazards plan that is designed to respond to many types of events, rather than having a separate plan for an earthquake, another plan for a fire, another for a tornado, and so on. An all-hazards plan will include many parts of the local healthcare system, including hospitals, long-term care facilities, mental health facilities, emergency medical services, medical laboratories, public health departments, and individual health providers.

When a community needs help from neighboring communities, state or regional agencies, or federal resources, it should expect that help will not arrive for 72 hours. It takes time for specialized teams located around the country to be notified, gather what they

A Closer Look

Emergency Medical Services

An Emergency Medical Service (EMS) is a local organization that provides emergency medical care in the field and rapid transportation to a hospital-based emergency department. This care is provided by emergency medical technicians (EMTs) and paramedics who follow protocols dictated and supervised by a physician. EMTs are trained in ambulance operations and procedures. Squads are sent out from an EMS dispatch center, such as 911, that receives the calls for help. EMS has different names across the United States, including emergency, first aid, safety or rescue squad, or ambulance corps.

need, and travel to the affected area. Because of the time lag, the local community must be prepared to handle the initial response.

National Incident Management System

In 2003, the U.S. Department of Homeland Security mandated that all governmental and nongovernmental entities follow new federal guidelines called the **National Incident Management System (NIMS)**, to prepare for and respond to emergencies. A local agency or organization must adopt the NIMS in order to receive federal dollars for emergency preparedness.

The NIMS is a comprehensive national approach to managing all types of emergencies. It provides a national template that enables all levels of government, the private sector, and nongovernmental organizations to work together efficiently and effectively to manage emergencies regardless of cause, size, location, or complexity.

The NIMS covers incident management and emergency prevention, mitigation, preparedness, response, and recovery programs and activities. The system:

- Establishes common principles so that every responder will understand how the affected community deals with an emergency.

- Establishes standard procedures for managing resources and facilitating coordination.
- Is flexible in scale so that it may be used for all incidents, small or large.

Incident Command Systems

Many agencies, institutions, and professionals must work together at the scene of an emergency. The local community agency sets up a special system to organize and coordinate response at the scene. This system is called an **Incident Command System (ICS)**. An ICS is a flexible management system for organizing and coordinating emergency response. It can be used in all types of emergencies, and everyone will know what to do.

The National Incident Management System, local governmental agencies, and hospitals all use incident command systems. These systems provide a configuration and a common language that help different agencies, institutions, and professionals work together.

Clear Authority and Roles

An ICS has a top-down structure. A single person is in charge. The **incident commander**, at first the most senior person to arrive on-site, assumes control of the response and assigns responsibilities. The incident commander communicates the status of the response to the emergency operations center. Command is usually transferred to more senior officials as they arrive.

If the event is very complex and requires a variety of skills and expertise, several professionals may direct the response together. Each is in charge of a particular workforce. This approach is called a unified command. In a **unified command**, all agencies with responsibility for the incident work together to establish a single, coordinated plan. Together they decide on a common set of objectives, plan jointly for operational activities, and maximize resources.

An ICS includes a **chain of command**, which clarifies reporting relationships and eliminates the confusion caused by multiple,

An incident command briefing at the FEMA Emergency Operations Center was set up in response to the disastrous tornado that struck Greensburg, Kansas, in May 2007. Briefings were held twice daily with representatives of relief organizations and town, county, state, and federal governments.

conflicting directives. The incident commander or commanders dispatch **emergency response providers** to the scene, usually through a community 911 system. Trained for their special responsibilities in advance, emergency response providers include public safety workers, police officers, fire fighters, and emergency medical technicians and paramedics.

Task Groups

The leaders of an Incident Command System organize the emergency responders into five groups: command and the four reporting groups: operations, planning, logistics, and finance and administration. Regardless of whether the emergency is routine or catastrophic, these five groups always make up an ICS.

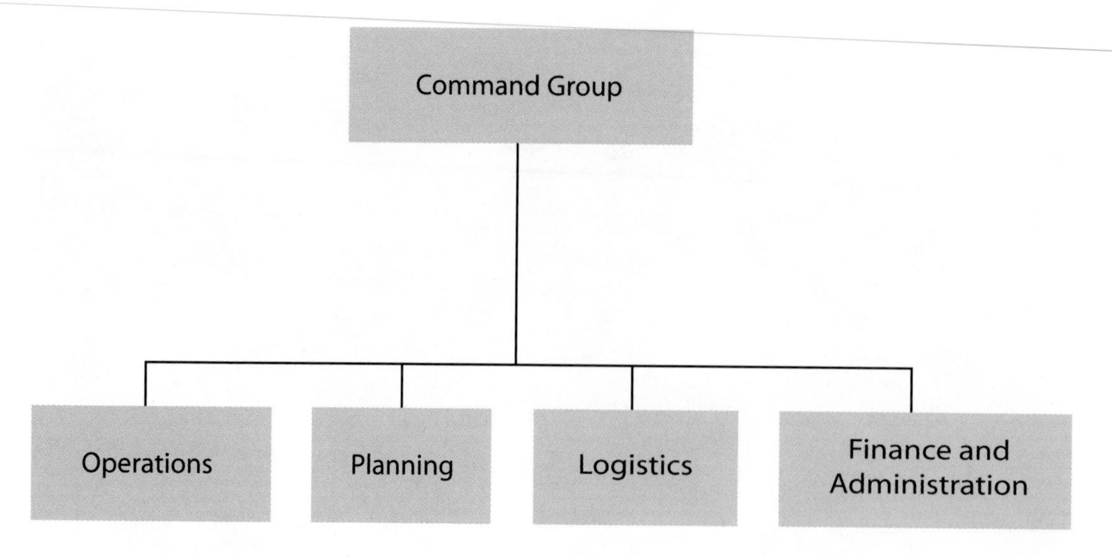

General organization of an Incident Command System (ICS)

The five task groups and their functions are:

- Command Group—Sets the goals and objectives for the response. An example of a command is: "Put out the fire and rescue any people who are in the building." The Command Group is led by the incident commander.
- Operations Group—Carries out directives necessary to achieve the objectives, for example, caring for victims of a fire or setting up safety zones around the building.
- Planning Group—Gathers, processes, and documents information about the response and communicates it to the leaders.
- Logistics Group—Provides transportation, supplies, equipment maintenance, fuel, food services, communications, information technology support, and medical services for emergency responders.
- Finance and Administration Group—Tracks costs and addresses issues such as reimbursement and claims.

A Closer Look

Incident Command System in Action

In 1993, a terrorist bomb exploded in the parking garage of the World Trade Center. Six people were killed and more than 1,000 were injured. New York City put an Incident Command System into effect immediately. The system continued for 23 days. A command post was established, and requests for interagency resources were made through the incident commander. Victim evaluation, treatment, and transport sectors were identified. A staging area was created. Following the initial setup, public-safety measures were taken and other services were put in place. Controlling traffic was critical so that supplies could be brought in and distributed. The police department closed the Brooklyn Battery Tunnel to keep access routes open for rescue vehicles. The Incident Command System allowed different agencies and responders to work smoothly together.

National Emergency Response

Emergencies that require federal assistance are complicated and require special planning. Federal resources must be formally requested. If approved, local officials must coordinate with personnel coming from across the United States to manage the extra supplies and equipment. The local agency remains in charge even after outside resources arrive. It also must reimburse the federal government for a portion of the resources that are sent.

FEMA

The **Federal Emergency Management Agency (FEMA)** is part of the U.S. Department of Homeland Security. FEMA is the lead federal agency during any national emergency. It has the primary responsibility for coordinating the regional response and providing management and response staff. FEMA also is the lead agency in charge of coordinating the provision of shelter, food, and first aid to victims.

A request for federal assistance begins with the Local Emergency Management Agency. The head of the local agency contacts his or her counterpart in the state government, asking for federal assistance. The Governor, committing state funds and resources, makes a request to the President that a federal disaster be declared. This is known as a **Presidential Declaration**.

FEMA determines whether the situation qualifies for federal aid under the Robert T. Stafford Disaster Relief and Emergency Assistance Act. The Stafford Act provides federal support to state and local governments so that they can carry out their responsibilities during disasters. Federal resources can include financial assistance, personnel, medicines, food, and other consumables.

The request for outside help can come either after a disaster has hit or in anticipation of impending disaster. When a disaster is imminent, the decision can be made to have supplies and personnel ready, strategically placing them nearby so that the response can reach the area more quickly. The goal is to lessen or avert a catastrophe.

National Response Framework

The Department of Homeland Security (DHS) developed the **National Response Framework (NRF)** to ensure a timely and effective federal response when a Presidential Disaster Declaration is made. While written for senior officials, the NRF can inform everyone who responds to an emergency.

This framework is a guide for coordination by communities, governmental agencies, the private sector and non-governmental partners. The NRF ensures that key roles and responsibilities in emergency response are aligned across the United States. The NRF emphasizes both employer and personal preparedness. It provides guidelines so that various entities can organize, operate and interact with one another in a unified manner.

It establishes a layered response. States have the primary responsibility for protecting their citizens, but all levels of government are linked and ready to help. When needed, a community's emergency operations plan should be flexible, adaptable, and **scalable**, or able to change the size of the response from small to large. Each level of government is expected to adapt and apply the general principles of the NRF.

The National Disaster Medical System

The federal government has also set up the **National Disaster Medical System (NDMS)**, which is responsible for sending medical teams, equipment, and supplies to a disaster area. It also transports ill and injured patients from the disaster to a safe area and provides care for patients in hospitals away from the disaster site.

The NDMS includes more than 9,000 medical, mortuary, veterinary, pharmacy, and nursing personnel organized into 107 teams across the nation. More than 1,800 U.S. hospitals are registered with NDMS to receive and care for patients.

The National Disaster Medical System includes various specialized teams. All team members are registered and trained in advance. They are volunteers that participate in meetings, training sessions, and drills and who are paid by the federal government if they are sent to a disaster site. Three of the specialized teams are:

- **Disaster Medical Assistance Teams (DMAT)** provide rapid response that supplements local medical resources. The team of medical professionals and support staff often sets up operations in a temporary medical site. There are also specialized teams for special medical conditions such as burns and mental health emergencies.
- **National Pharmacy Response Teams (NPRT)** provide mass vaccinations or pharmaceutical treatments for hundreds of thousands of people. Pharmacists and pharmacy technicians staff these teams.
- **Disaster Mortuary Operations Response Teams (DMORT)** provide victim identification and mortuary services. DMORT team members include funeral directors, medical examiners, pathologists, medical records technicians, dental assistants, radiographers, mental health specialists (to help people whose family members have died), and support staff.

Medical Facility Emergency Management

Hospitals and other healthcare facilities are also involved in comprehensive emergency management. To receive accreditation, hospitals and long-term care facilities are required to meet specific standards of emergency preparedness. Hospitals that are accredited by their state must follow state regulations for emergency planning that may include special drills and general procedures. Hospitals and health clinics that receive federal money, for example from Medicare, must follow federal guidelines.

Hospital Emergency Operations Plans

The Joint Commission, which accredits healthcare facilities, requires that all hospitals, long-term care facilities, behavioral

Members of Massachusetts Disaster Medical Assistance Teams (DMAT) attend a briefing to assist the treatment of patients who were arriving at a hospital after Hurricane Dennis struck Pensacola, Florida, in July 2005.

health facilities, and ambulatory care facilities have detailed emergency operations plans. The plans must be scalable so that the hospital could respond to an emergency such as a small fire or a disaster that devastates a community.

A **hospital emergency operations plan** covers mitigation, preparedness, response, and recovery. It encompasses concerns such as:

- How, when, and by whom the response and recovery phases of the plan are to be activated.
- How staff members will be notified when they are needed.
- Staff assignments.
- Procedures for providing patient care.
- Procedures for obtaining critical supplies and medications.
- Roles of support and security staff.
- How communication with the media is to be handled.
- Plans for evacuating the building and establishing care at an alternative site.
- Procedures for notifying the LEMA if the plan has been activated.

The Joint Commission requires that hospitals practice their emergency plans twice a year to test the plans and ensure their effectiveness. These tests can be either a response to an actual emergency or a planned drill. Drills must include every staff member in every building that is used by the hospital. Drills should test the communication systems used by the hospital, coordination, and incident command structures. Additionally, if hospitals have an emergency department, at least one of the annual drills must include volunteers who act as patients in the emergency department.

In the event of an emergency, it is important for every healthcare worker to be familiar with the following:

- Personal responsibilities during an emergency.
- What the facility plan expects of a department.

- Whom workers report to and the chain of command.
- The codes used by the facility to signal that there is an emergency (such as four bells).
- How supervisors will notify people when the plan is activated.

If you are notified of an emergency, go to your assigned location as soon as possible. You will likely be asked to remain on duty or be reassigned. If you are notified of an emergency when you are at home, report to work at the usual time, unless you are told to do otherwise.

Hospital Incident Command System

The Joint Commission and federal guidelines require hospitals to follow a **Hospital Incident Command System (HICS)** when responding to emergencies and disasters. Like other incident command systems, the hospital system clarifies roles and responsibilities. The written responsibilities are defined and listed in advance on **job action sheets**. The incident commander is often the chief executive or chief operating officer of the hospital. The system enables facilities to handle an emergency without delay by establishing a chain of command and specifying staff responsibilities. Using an Incident Command System helps hospitals work smoothly with other responding agencies because all will use the same terminology and organizing principles.

Medical Office Emergency Plans

Many private medical offices and clinics have an emergency plan. The office manager often creates the plan. Medical office plans should contain procedures for evacuating the building, including how all will be accounted for, and a central meeting place. Evacuation plans should include procedures for shutting down critical operations, such as disabling any central oxygen or medical gas system. Protection of health records is another consideration. The emergency plan should include procedures for evacuating patients with disabilities and limited mobility.

Medical offices should plan to provide care within the emergency planning guidelines established by community and regional healthcare systems. Coordination is also important between the office and any affiliated and regional hospitals and the local health department.

The plan should include procedures for notifying staff of an emergency. Telephone lists and trees are a common way to notify all staff quickly. The plan should also have a contact number outside the region, so employees can check in, if necessary.

Medical offices should also educate their patients, especially those with chronic illnesses, about emergency preparedness. Private health providers play a major role in ensuring continuity of care during a disaster.

Other Health Facility Disaster Plans

Long-term care, behavioral healthcare, public health clinics, and other healthcare facilities have emergency plans. It is important for these facilities to remain open during an emergency because regular patients will need to continue receiving their customary healthcare, and many others affected by the emergency may also seek care.

Staff should receive training based on the plan. Drills should be conducted semiannually. The EOP should provide for an emergency operations center and the activation of an ICS during an emergency. Centers should conduct inventories of emergency equipment and supplies and arrange for replacements if needed. Finally, it is important for all health facilities to be integrated into the community response plan.

Summary

Emergency management has four phases: mitigation, preparedness, response, and recovery. All disasters are managed locally, and communities have their own emergency management agencies. Additionally, all communities follow national guidelines. These guidelines provide uniformity in methodology and terminology, which makes it possible for people responding to emergencies to operate effectively and efficiently. Procedures involving Incident Command Systems establish clear lines of authority. The federal government becomes involved in emergency response at the request of the governor, initiated by the LEMA. FEMA is the lead agency at the national level. Hospitals also have emergency operations plans and Incident Command Systems, while small private practices and other healthcare facilities need to establish their own plans for dealing with emergencies.

Key Terms

- chain of command
- comprehensive emergency management
- Disaster Medical Assistance Team (DMAT)
- Disaster Mortuary Operations Response Team (DMORT)
- emergency operations center (EOC)
- emergency operations plan (EOP)
- emergency response providers
- Federal Emergency Management Agency (FEMA)
- hospital emergency operations plan
- Hospital Incident Command System
- Incident Command System
- job action sheet
- Local Emergency Management Agency (LEMA)
- mitigation
- mutual aid
- National Disaster Medical System (NDMS)
- National Incident Management System (NIMS)
- National Pharmacy Response Team (NPRT)
- National Response Framework (NRF)
- preparedness
- Presidential Declaration
- recovery
- response
- scalable
- unified command

Reinforcing Terminology

Choose the word or phrase from the list of key terms that best matches each of the following descriptions.

1. Set up by the federal government to send medical teams, equipment, and supplies to a disaster area ________
2. Establishes uniform procedures for responding to disasters across the country ________
3. Maintains primary responsibility for emergency management and response ________
4. Provides a written record of hospital personnel roles and responsibilities in an emergency ________
5. Establishes that communities will assist each other in responding to or recovering from a disaster ________
6. Can administer mass vaccinations to hundreds of thousands of people ________
7. Needed to receive federal assistance for disaster response ________
8. Establishes methods on how federal departments and local agencies will work together ________
9. The lead federal agency in a disaster response________
10. Needed by hospitals to be accredited by the Joint Commission ________

Reviewing Concepts

Choose the response that best answers the question or completes the sentence.

1. Which phase of comprehensive emergency management involves creating evacuation plans?
   a. mitigation
   b. preparedness
   c. response
   d. recovery

2. Transporting flood victims to safe locations is an example of
   a. mitigation.
   b. preparedness.
   c. response.
   d. recovery.

3. A community might install a siren system to minimize hazards resulting from a
   a. tornado.
   b. winter storm.
   c. heat wave.
   d. terrorist attack.

4. The part of an emergency operations plan that clarifies reporting roles is the
   a. unified command.
   b. incident command.
   c. direct command.
   d. chain of command.

5. In what percent of emergencies is the federal government involved?
   a. 1 percent
   b. 10 percent
   c. 50 percent
   d. 100 percent

6. During a disaster response, where would you expect to find the incident commander?
   a. at the scene of the disaster
   b. at the emergency operations center
   c. at police headquarters
   d. at the state capitol

7. In an Incident Command System, which group provides transportation and supplies for emergency response providers?
   a. planning
   b. operations
   c. logistics
   d. finance/administration

8. Who must originate a request for federal disaster relief?
   a. congressional representative
   b. governor of state
   c. head of FEMA
   d. head of LEMA

9. Which team belonging to the National Disaster Medical System provides victim-identification services?
   a. Disaster Medical Assistance Team
   b. National Pharmacy Response Team
   c. Disaster Mortuary Operations Response Team
   d. Federal Emergency Management Team

10. How often should healthcare facilities conduct emergency preparedness drills?
   a. monthly
   b. once a year
   c. twice a year
   d. every other year

Exploring Emergency Preparedness Issues

1. Go to *www.fema.gov/news/disasters.fema,* the 2008 Federal (Presidential) Disaster Declarations, to see the types of disasters that occurred during the past five years in your region of the country. Summarize this information in a one-page report.
2. Contact the Local Emergency Management Agency in your community. What types of emergencies are they planning for? Are they planning for more than one kind of disaster or are they using an all-hazards approach? Does their plan include healthcare facilities where you work or might work? How are they included? Discuss your responses to these questions in one or two pages.
3. Contact a local hospital and ask how to volunteer to be a victim in a drill. If time allows and you are able to volunteer, write a two-page report about your observations and the types of activities your profession carried out in the drill. If you noticed that the drill identified a weakness in the response, include that in your report (as well as your recommendation for improving the emergency response).

CHAPTER 3

Personal Preparedness

What you will learn:

- What you need to know and how you can apply your professional skills to emergency situations
- Your role and responsibilities in your employer's emergency action plan and disaster training procedures
- What is entailed in volunteering for emergency and disaster work
- Recognizing the sources and signs of stress and learning how to cope with stress
- Information on emergency preparedness for yourself, your family, and your neighborhood

Emergency preparedness is for everyone, including government agencies, hospitals, and you. You can prepare for your part in emergency preparedness and response by developing related skills, learning the emergency plan of your employer, exploring volunteer opportunities, finding ways to cope with the stress of response work, and preparing your family for whatever may happen.

Sharpening Your Response Skills

As a health professional, you have many clinical and administrative skills that are invaluable in emergency response. You have basic and specialized medical skills, as well as the ability and background to fill many important support roles.

Professional Skills

All allied health professionals have special skills related to emergency response. For example, respiratory therapists can provide airway management and ventilator support, which would be particularly valuable in responding to disasters involving chemicals or pandemic flu.

In emergency response, allied health professionals use their skills in new and more intense contexts. However, they should always stay within their legally and professionally defined scope of practice. They should only perform tasks that they have been trained and qualified to do. For example, if an emergency response requires a large volume of imaging or lab tests, a respiratory therapist could not jump into the role of radiographer or lab technologist. Many health professionals feel more confident about their potential roles in a disaster when they realize that they will not be asked to perform tasks for which they have not been trained.

The final chapter of this book explores the special response skills of a variety of allied health professions. Meanwhile, as you learn about emergency procedures, keep an "I can" list, linking your professional skills to those that are needed in emergency response. The list will probably be much longer than you expect!

Basic Medical Skills

Basic medical skills are part of the education of most health professionals. These skills include taking vital signs, providing first aid and cardiopulmonary resuscitation (CPR), taking patient

Abbreviations

CERT	Community Emergency Response Team
CISD	critical incident stress debriefing
CPR	cardiopulmonary resuscitation
EMAC	Emergency Management Assistance Compact
ESAR-VHP	Emergency Systems for the Advance Registration of Volunteer Health Professionals
IV	intravenous
MRC	Medical Reserve Corps
OSHA	Occupational Safety and Health Administration

histories, and drawing blood. Although you may not use these skills in your daily work, you may reasonably be expected to apply them in an emergency. Disaster response requires many workers with these basic medical skills. Practice these skills often to keep them sharp.

First-Aid Training

Imagine that you are the first person to arrive at the scene of an emergency and people are injured. Would you be able to provide first aid until more help arrived? **First aid** is simple, potentially life saving medical techniques delivered with minimal or no medical equipment. Providing first aid is something that everyone can do with the right training.

First aid is part of most educational programs in allied health. Many healthcare organizations require that healthcare professionals keep their certification in first aid current by taking courses through national and local groups. Local chapters of the American Red Cross teach a variety of courses. They also offer training in which the first portion is completed online. You can then work in person with a Red Cross instructor to

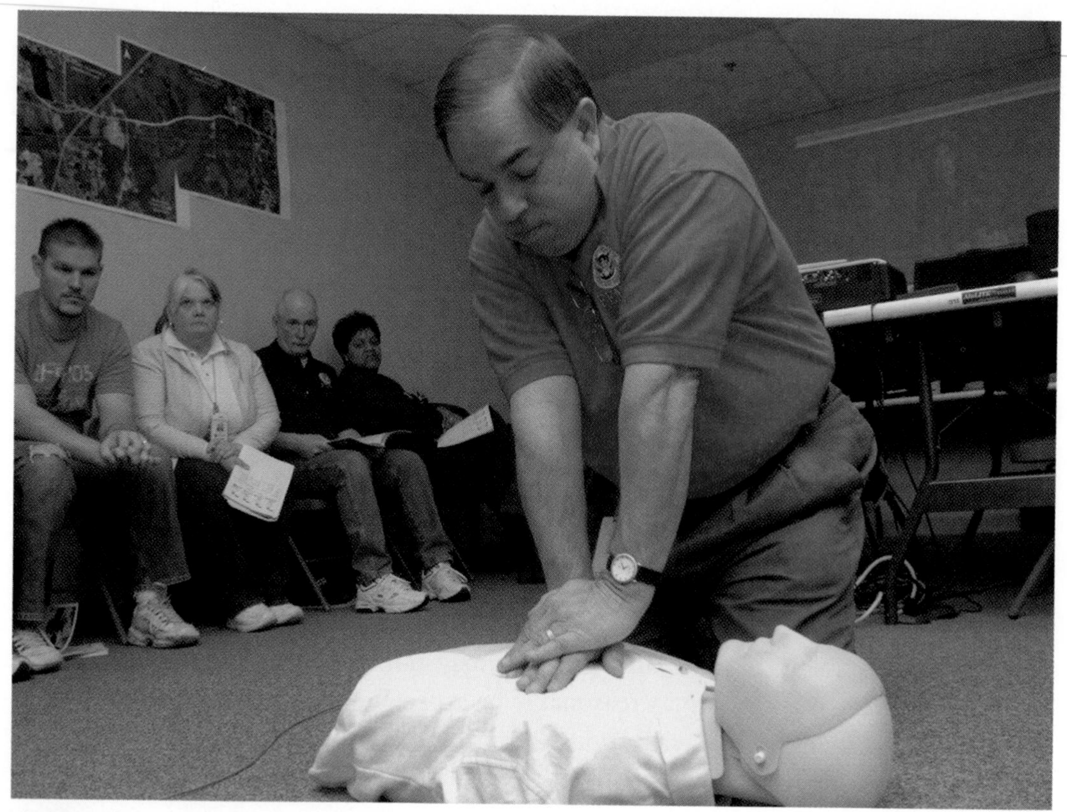

Continual training and practice are important in emergency preparedness. Here, CPR techniques are being practiced in a Red Cross training class.

practice and polish your skills. Many hospitals also offer first aid courses.

Working without Technology

Could you perform the essential tasks of your job without the technology that is typically used? What would you do if the power goes out and backup power fails? Computers, intravenous (IV) pumps, or respirators require electricity to operate. Learning the low-tech ways to perform core tasks is an important part of emergency preparedness for all health professionals.

Many instructors and professionals who have been in their fields for many years know how the work was done before technology. They can teach you how to work without technology. If this is not done as a matter of course and if you work with seasoned professionals, ask them to help you learn.

Support Skills

Allied health professionals can fill many support roles well in emergency response because of their experience in working with patients and the healthcare system. When there are many patients and little time, support workers are critical to effective emergency response. For example, at centers that distribute mass immunizations or medications, volunteers are needed to translate, maintain stores of medicines and supplies, direct patients, provide communications links, control crowds, and more.

Support work takes many forms. For example, in the hours that followed the attack of the World Trade Center in 2001, sonographers (ultrasound specialists) at White Plains Hospital in New York State, a community hospital 30 miles north of the World Trade Center, were asked to prepare rooms so that families of victims who lived nearby could view the bodies. The sonographers cleared rooms, got gurneys ready, and had food and drink available for families. This work was important to the families and the medical facility. These health professionals drew on their experience working with patients and their facility, and performed an important service.

Employee Preparedness

Most healthcare professionals respond to emergencies as employees. What tasks they are expected to perform and how they are supposed to do the tasks are detailed in the emergency plan of their employer. Become familiar with the plan of your employer so that you can respond effectively and safely.

Employer's Emergency Action Plan

In Chapter 2, you learned about the emergency management plans that every hospital and healthcare facility must have. The U.S. Occupational Safety and Health Administration (OSHA)

A Closer Look

A Good Emergency Action Plan

A good plan should identify and explain the actions to be taken in an emergency. A good emergency plan:

- Identifies potential threats and hazards.
- Identifies emergency response and reporting procedures.
- Describes how to report an emergency.
- Defines roles and responsibilities of employees.
- Describes procedures for warnings and communication.
- Identifies the location and explains the use of emergency equipment.
- Describes emergency shutdown procedures.
- Describes evacuation procedures.
- Provides exit maps or diagrams.
- Tells how all personnel will be accounted for.
- Describes procedures for taking shelter in the facility.
- Identifies rescue and medical tasks.
- Provides for training.

requires that all companies, including healthcare, with more than 10 employees have a written emergency action plan.

Employees learn about the emergency action plan through orientation sessions when they are hired, at departmental meetings, and during emergency drills. Any employee can ask to see the plan. The plan will include the assigned tasks your department will be responsible for in various types of emergencies. For example, if your work area contains special hazards, such as flammable or radioactive materials, you may be responsible for securing these materials in an emergency. The plan will also explain how you will be notified when an emergency has occurred.

Disaster Training

Emergency drills are simulated emergencies that allow people to practice putting an emergency plan into action. Health professionals may participate in drills conducted by their employers or a local emergency management agency. Drills help train response workers. They also identify strengths and weaknesses of

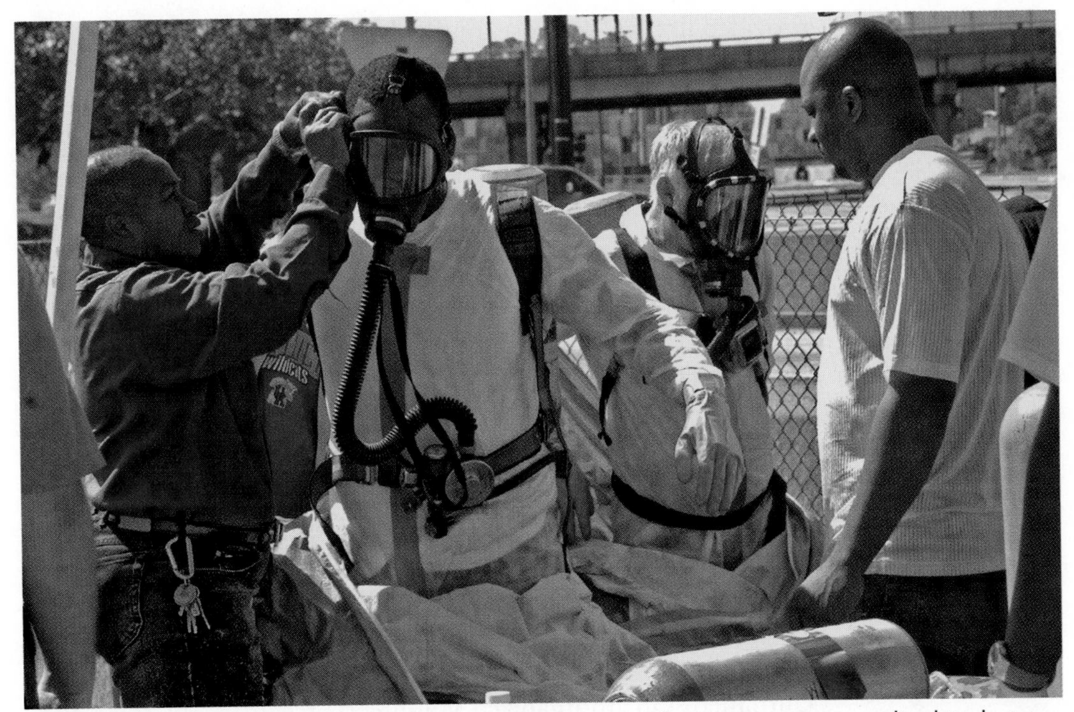

A hazmat instructor (*left*) adjusts the respirator mask of an emergency medical technician team member. The team members are preparing to respond to a simulated chemical spill and rescue operation.

an emergency plan. After a drill, everyone should help evaluate how things went and recommend improvements.

Employees may also receive training in emergency procedures or equipment. If you do not receive training, look for training opportunities such as online classes or training at local response organizations.

How to Volunteer

During the devastating 2005 hurricane season in the Gulf States, more than 50,000 health professionals from across the country volunteered to help. They were deployed through state agencies, the American Red Cross, or the federal government. Healthcare volunteers filled in at local hospitals, assisted at local shelters, provided first aid to those injured by the storms, assisted in health assessments, and staffed special needs shelters, community health centers, and clinics.

While responsibilities to the employer come first, many healthcare professionals also want to volunteer to help in their communities, other states, or other countries. Allied health professionals can get involved in many ways. They can make a real difference for patients and communities.

Preregistration

When a disaster takes place, most allied health professionals are eager to help. However, health professionals who want to serve as volunteers must register in advance. By doing this, they join an organized effort. They will know details of the community plans and the types of tasks they might perform during an emergency. They will receive training and be assigned to a team that has access to equipment and supplies and support for their personal needs. Untrained and unorganized volunteers, regardless of their good intentions, create problems during a disaster. Their competencies are unknown, and there may be insufficient food or housing for them.

Volunteer Sites

As a volunteer, you may work in another hospital to support local staff by performing tasks you are professionally trained and licensed to do. You may also work at an alternative care site, especially if hospital facilities have been damaged or destroyed.

Before volunteering to work in a disaster area, know what to expect. Disaster areas usually have limited supplies of food, safe water, and medicine. Responders may need current immunizations for diseases such as hepatitis B and tetanus. Workers often have long shifts with little rest, so physical stamina and health are important considerations. They may also need transportation and special clothing and equipment to get to and stay in the affected region. To care for themselves over an extended period of time, volunteers are expected to bring their own clothing, hygiene products, and so on.

A Closer Look

Expected Working Conditions

Health professionals who volunteer for a federal assignment in a disaster area should expect to:

- Work 12-hour shifts for two weeks or longer.
- Be housed in tents, possibly with no showers, and sleep on bed rolls or cots.
- Not have air-conditioning.
- Stand for long stretches of time.
- Eat military "ready-to-eat" meals.
- Use portable toilets.

Volunteers must also be psychologically prepared. They will be caring for survivors who may have lost family members and homes. Victims may have witnessed violence or suffering and may be feeling anguish, anger, and sadness. Volunteers have to be able to provide support, while experiencing many of the same emotions.

Credentialing and Liability

All health volunteers have to be **credentialed** to verify their identity, license, training, skills, and competencies. Their credentials determine which tasks they are permitted to perform in an emergency. Only those who are registered in advance can be sure that their credentials will be verified so that their skills can be used where they are needed most.

Many health professions have state or national licensing. If they are preregistered and credentialed to volunteer, then their licensure, certification, and credentialing can be temporarily transferred from one state to another when an emergency has been declared. Otherwise, health professionals may not be permitted to work in a different state.

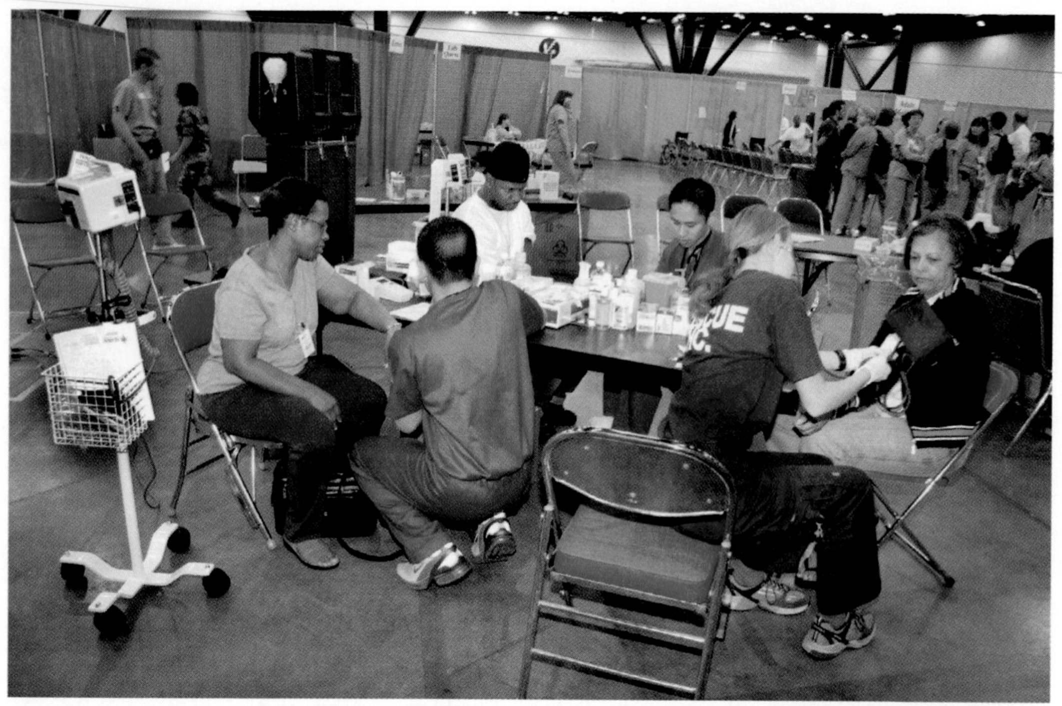

Medical volunteers give blood pressure tests to residents of the George R. Brown Convention Center in Houston, Texas, in September 2005. Hundreds of displaced New Orleans residents were housed in the convention center after Hurricane Katrina.

Volunteers must be registered with a response agency to have liability protection. The extent of the liability coverage depends on who deploys the volunteers. Volunteers on federal response teams are provided with professional liability coverage when on federal assignment.

Many health professionals are covered for liability by the organizations for which they work and do not have independent coverage for malpractice. Before considering responding to a disaster, healthcare workers should check the Good Samaritan laws of their state to see if the statutes provide any protection to those providing emergency care during a disaster. Additionally, if healthcare workers have individual liability insurance coverage, they need to determine whether the coverage is effective if they work outside their usual practice settings.

Training

Preregistered volunteers may receive advance training for the tasks they may be asked to perform. Sometimes qualified people may be unavailable at the time of the emergency. As a result, agencies are developing **just-in-time training**, which is specific last-minute training for tasks the volunteer will be performing on the scene.

Where to Register

To volunteer for disaster work in your state, contact the emergency preparedness office of your state or local health department. You will be directed to the person who handles the recruitment of volunteers. At registration, indicate whether you are willing to serve only in your community or whether you are available to work in other areas.

Each state has an **Emergency System for the Advance Registration of Volunteer Health Professionals (ESAR-VHP)**. These systems are designed to assess and preregister volunteer health professionals for work within their own states and in other states during public health and other emergencies. Each state-based system contains readily available, verifiable, current information about the identity, licensing, and credentials of a volunteer.

People who preregister can be deployed through the **Emergency Management Assistance Compact (EMAC)**. EMAC is a mutual aid agreement that facilitates the sharing of resources, personnel, and equipment across state lines during disasters. It covers all states, the District of Columbia, Puerto Rico, and the U.S. Virgin Islands. EMAC was used to deploy more than 2,000 healthcare professionals from 28 states to areas affected by Hurricanes Katrina and Rita.

At the federal level, the **Medical Reserve Corps (MRC)** coordinates with communities in the deployment of healthcare volunteers. MRC units are organized by local communities to enhance the emergency and public health resources that are already in place. All local MRC unit leaders are asked to ensure that local community needs are met before volunteers are sent out of the area. Workers who sign up for the MRC are placed on a list of providers who are preapproved to respond in an emergency. The MRC will verify credentials in advance and will give the workers an identification card. The local or state department of health or the office of emergency management maintains the list of credentialed people in the MRC. Go to the national MRC website at *www.medicalreservecorps.gov/AboutVolunteering* and use the link to locate the MRC unit in your state.

The National Disaster Medical System (NDMS), introduced in Chapter 2, is composed of rapid-response teams of healthcare professionals and logistical and administrative personnel. The Disaster Medical Assistance Teams (DMAT), also introduced in Chapter 2, evaluate patients, provide basic medical care, and prepare patients for evacuation in mass casualty incidents. DMATs also provide primary healthcare or assist overloaded healthcare workers. Other teams provide pharmacy or mortuary services. To volunteer with a National Disaster Medical System team, go to *www.hhs.gov/aspr/opeo/ndms/index.html* or call 1-800-USA-NDMS.

Nonmedical Opportunities

You can volunteer to provide support services in emergency response. For example, you can volunteer with the American Red Cross, which offers shelters, provides food, and raises money to aid victims. Many communities have local Red Cross chapters that train and organize volunteers.

Another volunteer opportunity is a **Community Emergency Response Team (CERT)**. The CERT program educates people about disaster preparedness and trains them in basic disaster response skills, such as fire safety, light search and rescue, and disaster medical operations. To become a CERT member, people must take CERT training from a sponsoring agency such as an emergency management agency, fire department, or police department. Locate a program in your community through the CERT website, *www.citizencorps.gov/*.

To find other nonhealthcare volunteer opportunities, go to USA Freedom Corps at *http://usafreedomcorps.gov/*.

A Red Cross volunteer loads supplies onto a pickup truck of a local resident affected by flooding in Ohio in August 2007. The services of volunteer agencies are very important during responses to natural and man-made disasters.

Preparing for Stress

Regardless of the emergency, wherever you respond, you will probably encounter situations that cause stress. Some sources of stress cannot be avoided, but there are many ways to reduce and cope with stress. Personal preparedness includes anticipating that you will experience stress and knowing how to deal with it.

Sources of Stress

Everyone works under difficult conditions in a disaster. Even with the best planning, situations may be chaotic. Health professionals may feel that no matter how much they do, it is not enough. Some victims will be grateful for the help they receive, while others may be angry at the situation and take their anger out on those who

A Closer Look

Possible Sources of Stress

During a disaster, stress can come from many sources, including:

- You, your family, friends, or coworkers may be among the victims.
- Communications may be down, so it may not be clear what is happening.
- Supplies may be inadequate.
- Weather conditions may be extreme.
- Tasks are likely to be physically exhausting.
- Some work will be dangerous.
- You may become sleep-deprived.
- You may feel that you cannot do enough to help victims.
- You may witness suffering and death.
- There may be struggles over decision-making and authority.
- The disaster may bring back painful memories of previous experiences.

are helping them. Healthcare professionals may witness people dying violently, or they may have to handle the remains of the dead. Additionally, workers may be victims. Their homes may be destroyed, or members of their families may be hurt or worse. It is normal and expected that people will experience stress under these circumstances. Even experienced emergency responders are affected.

On the positive side, many people find that working together for a common good brings them closer together. The experience can influence professional and personal beliefs or priorities and provide a deep sense of satisfaction for making an important contribution. Even with this positive aspect, expect the work to be difficult and stressful.

Recognizing Signs of Stress and Burnout

No one who sees or experiences a disaster can remain untouched by it. While most healthcare professionals function well during and after a disaster, it is normal for them to feel stressed or saddened about what has happened—to them, the people they care about, and their community. Many people feel angry, powerless, frightened, or guilty for surviving. These are normal reactions.

During the response, everyone concentrates on helping the victims of the disaster. The needs of the workers may be initially ignored or seen as less important than those of the victims. Disaster workers may not be prepared for their own emotional reactions to the situation.

Sometimes stress builds up and leads to **burnout**, which is a condition in which people suffer emotional exhaustion. They may no longer feel enthusiastic about their work, they may work less efficiently, and they may feel tired even when they get enough sleep. They may be irritable about their jobs and toward the people they work with. Other warning signs include having difficulty concentrating or sleeping or a variety of physical symptoms such as headaches or gastrointestinal upsets. People experiencing burnout may drink too much coffee or alcohol or smoke too much. When workers burn out, they do not work as hard, which diminishes the efficiency of their team. They may put themselves in physical or emotional danger if they are not concentrating on what they are doing while working.

People who work at the site of disasters may face more stress when they return home. What they experienced in the field is probably different from their normal routine. When they return home, they may need time to adjust. However, their family will probably want their attention or help after their absence. Family members may be angry that the worker was gone, especially if they were also victims of the disaster.

Unlike emergency workers whose full-time job involves dealing with crises in the community, healthcare professionals often do not have much experience in dealing with disasters. It will help you to know in advance the ways you may react. Learn the signs of stress and burnout in yourself and your coworkers. Some of these signs include:

- *Cognitive reactions.* Stress can affect the ability to reason or think clearly. People may be unable to concentrate, or they feel confused. Sometimes memories of the disaster intrude on their consciousness without warning.

- *Emotional reactions.* People may feel angry, shocked, or frightened. They may have a hard time accepting what happened. They may feel guilty because they are okay, while others are not. People who have gone through disasters feel grief, whether for people they know or for victims they do not know. They may feel despair and a sense of helplessness. They may also seem irritable and short-tempered.
- *Social reactions.* Stress may make people feel distant from those they are normally close to. Some may pull away from family and friends. Often stress will put a strain on personal relationships. Some people who experience stress may turn to drugs or alcohol as a way to cope.
- *Physical reactions.* People may be tired, even when they get enough sleep. Or, they may be overexcited. People often experience insomnia or other sleep disturbances. Some people may have physical aches and pains that seem unrelated to a specific physical problem. Others may become more susceptible to illness than usual. Headaches and stomach aches are common symptoms of stress, as are a decrease in appetite and sexual desire.

Ways to Cope

Part of the responsibility of protecting workers from stress belongs to the disaster supervisors and incident commanders. These managers should recognize and show appreciation for the important work being done. They should be aware of the toll this work can take on people and arrange for adequate breaks during rescue and relief operations. They also should arrange for debriefing sessions at the appropriate times. These sessions provide an opportunity to discuss and share feelings. Organizations use several techniques to help disaster workers deal with stress. Look for and take advantage of opportunities to deal with stress, including:

- **Crisis intervention** is a set of techniques that helps people feel more in control during a crisis. In group or individual counseling, workers have an opportunity to share their feelings and help identify solutions.

- **Defusing** is a brief, informal procedure used to help disaster workers deal with their reactions to the events. In defusing, mental health professionals hold 10- to 30-minute sessions in which workers can express their thoughts and feelings about what they are doing. Workers may need to talk about an unexpected incident, such as an accident; the discovery of a disfigured body; or a conflict between workers, with a supervisor, or with a victim. Disaster workers, like victims, should be able to talk about their experiences and feelings in both informal and formal settings.
- **Critical incident stress debriefing (CISD)** is a technique used to help a group of disaster workers process their disaster experience. Most debriefing sessions last two to four hours. Critical incident stress debriefing helps people learn that they are not alone in their reactions to the events that they have experienced. Where debriefings are offered, participation is considered part of the job of a disaster worker. In addition to helping with dealing with stress on the job, debriefings can help workers develop a support plan for their return home.

You can use many techniques to reduce tension and anxiety during the response phase and in the weeks and months after you return home. Some of these techniques are:

- Recognize, understand, and appreciate your feelings. It is not abnormal or crazy to have strong feelings and reactions.
- Be tolerant of the reactions of other relief workers and victims. Disasters are a time of stress for everyone.
- Try relaxation techniques involving breathing, visualization, or muscle relaxation. A therapist can teach you these techniques.
- Talk to other workers about their feelings. Talking helps relieve stress and helps people realize that others feel the same way.

- Keep a journal.
- Take care of yourself. Get as much sleep as possible. Take rest breaks, eat properly, and avoid drinking large amounts of caffeine or alcohol. Physical activity, such as running, walking, or engaging in sports, is also helpful.
- If something especially upsetting happens, take a short break, use a relaxation exercise, or talk to someone about it. Do not dwell on what happened.
- If you have been working away from home, give yourself a few days to make the transition once you return. Tell your family that you need some time to yourself before you return to a normal schedule. Expect that other family members also will have been dealing with stress.
- If you experience prolonged symptoms, it is important to contact a therapist.

Preparing Your Family

Another aspect of personal preparedness happens at home. In times of emergency, if you do not know that your family is safe and secure, you may not be able to concentrate on your job or even come to work. To ease your mind, create your own emergency plan.

Family Emergency Plan

To create a family emergency plan, the first step is to identify the emergencies that are most likely to occur where you live. Then find out the recommended procedures for dealing with them. Most people know about and have some experience with the natural disasters that occur in their areas—hurricanes in Florida, earthquakes in California, and so on. However, other local hazards may be less obvious—a chemical plant, for example. Sources of information on local hazards include local emergency management agencies and Red Cross chapters. Ask for information on emergency procedures in the community, including how citizens are alerted. Emergency alerts may include warning sirens, television and radio announcements, automatic telephone dialing systems, and door-to-door notifications.

When you are familiar about the potential emergencies in your community, meet with the members of your family and create a plan together. Working as a team and sharing responsibilities will help all family members understand what they need to. Include children in some of the planning. Tell them that emergencies such as fire, severe weather, or whatever may occur where you live and that preparing for emergencies can make them safer.

In some disasters, government authorities will advise people to evacuate or to seek shelter in their homes. Be sure that everyone in the family knows that they should follow the instructions of the public agencies in charge of making such decisions. Explain that authorities make these decisions based on the type of emergency. For example, during a tornado warning, people are safer underground. In the case of a chemical release, an above-ground location is preferable because some chemicals are heavier than air and may seep into basements even if the windows are closed.

Communication during an emergency can be difficult. Families should decide ahead of time how they will contact one another, how they will get back together, and what they will do in different situations. All these actions should be written into the plan. Families should designate two meeting places—one right outside the home and another outside the neighborhood, such as

A Closer Look

Other Ways to Prepare Your Family

You can prepare for an emergency in other ways, including:

- Be sure that every family member knows how and when to call 911 or the emergency services number in your community.
- Take a first-aid course with your family. The American Red Cross and other community organizations offer courses.
- Check your homeowner insurance coverage.

A Closer Look

Family Contact Information

The following is an example of what everyone should carry in case of an emergency:

If I am called to work to respond to an emergency, our family has made the following arrangements so that all are cared for while I am on duty.

My child(ren) will be cared for by:

____
(Name)

____
(Phone Number)

____
(Address)

My pet(s) will be cared for by:

____
(Name)

____
(Phone Number)

____
(Address)

My parent(s) will be cared for by:

____
(Name)

____
(Phone Number)

____
(Address)

Our meeting place is ____________________________

We will all call ________________ at ________________
to report where we are and that we are safe. (Phone Number)

A Closer Look

Taking Cover

In a sudden emergency such as an earthquake, take immediate actions to protect yourself from injury. For example:

- Protect yourself from falling objects by going under a desk or table.
- Stand in a doorway.
- Move away from file cabinets or bookshelves that might fall.
- Face away from windows and glass.
- Move away from exterior walls.

a library, community center, or place of worship. Everyone in the family must know the address and phone number of these places. Emergency plans also should include family members leaving notes in designated places that tell others where they are going.

Families should ask an out-of-town friend or relative to serve as a "communication center" if family members become separated. Every member of the family should know who that person is, have the phone number readily accessible, and have a prepaid phone card to call the emergency contact. In case of evacuation, every member of the family should call or e-mail the out-of-town contact to tell this person where the family members are going. This is important because a disaster can interrupt or destroy the power and phone service, including cell phones. When people get to an area where they can call, they should contact the designated person. Evacuation planning should include special arrangements for family members who are ill or disabled. Be sure to have a plan to take care of pets. Most shelters for people do not allow pets.

All household members should keep a copy of the family disaster plan and emergency contact information in their wallets, briefcases, or backpacks. Contact information should include names, addresses, phone numbers, and where to meet in case of an emergency.

Once everyone knows what to do in an emergency, families should practice their plans, including a mock evacuation.

Neighborhood Planning

Schools, day-care centers, workplaces, and multistoried buildings should all have site-specific emergency plans. People should know about the places where their family members spend time and ask about emergency plans at those places. Know what is in the plan, and be sure it does not conflict with the procedures in your family plan. For example, parents with children should know whether the school will keep students at the school or send them home. Parents must also know how to authorize a friend or relative to pick up the children if they cannot. They should provide the school with up-to-date contact information for themselves and at least one other neighbor, relative, or friend.

Talk to your neighbors about how you can all work together in the event of an emergency. Some neighborhood organizations, such as a home association or crime watch group, may have disaster preparedness as part of their normal activities and will share the information with you.

Neighbors can decide who will check elderly or disabled neighbors. They can share contact information and make backup plans for children in case someone cannot get home.

Family Emergency Supplies

It is important that you and your family be prepared. Collect supplies and store them in a convenient location. Start small. Put together a **go bag**, which is a small collection of essentials that you and your family may need in the event of an evacuation. At a minimum, the go bag should contain the following:

- Copies of important documents in a waterproof and portable container. Include photocopies of insurance cards, passport, driver's license and/or other photo IDs, proof of address, bank account records, social security cards, and so on.
- Bottled water and nonperishable food such as energy or granola bars.
- An extra set of car and house keys.
- Credit and ATM cards and $50 to $100 in cash.

- Flashlight.
- Battery-powered AM/FM radio.
- Extra batteries for the flashlight and radio.
- A list of the prescription medications each member of your household takes, why they take them, and the dosages. If you decide to keep extra medications in the go bag, be sure to replace them before they expire.
- First-aid kit.
- Contact and meeting place information for the members of the household.
- A small map of the region.
- Child-care supplies and other special care items.

A go bag contains what you can quickly pick up and carry with you. You can pull together a more extensive collection of supplies to keep in your home. Every household should collect enough emergency supplies to take care of family members without outside help for at least three days. Everyone in the family should know that these supplies are for emergencies only. Once prepared, supplies should be updated and checked twice a year for the expiration dates of food and medications. Food and stored water should be replaced periodically to keep them fresh. A good time to schedule an update is when clocks are changed to and from daylight saving time.

The supplies to keep in your home should include everything in the go bag, plus the following:

- Water—have one gallon of water per person per day for at least three days for drinking and sanitation. You may want to store more if a family member is on medications that require water or increase thirst. Store water in plastic containers.
- At least a three-day supply of nonperishable food. Select foods that require no refrigeration, preparation, or cooking, and that need little or no water. Examples include energy, granola, or fruit bars; dried fruit; nuts; peanut butter; crackers; boxed juices; ready-to-eat foods, and foods for special dietary needs.

- Medications, both prescription and nonprescription. Be sure these do not expire.
- Unscented chlorine bleach and a medicine dropper. When diluted (nine parts water to one part bleach), bleach can be used as a disinfectant. In an emergency, sanitize water by using 16 drops of unscented chlorine liquid bleach per gallon. Water purification tablets or iodine tablets also work. Purchase these at stores that carry camping supplies.
- Manual can opener.
- Plastic sheeting and duct tape for **sheltering-in-place**, which refers to remaining in a facility or other designated, protective area, during a disaster.
- Moist towelettes, garbage bags, and plastic ties for personal sanitation.
- Basic tools (wrench, pliers, and so on).
- Pet food and extra water for pets.
- Sleeping bag or warm blanket for each person.
- Matches in a waterproof container.
- Mess kits, paper cups and plates, plastic utensils, and paper towels.
- Paper and pencils.
- Books, games, puzzles, or other activities for children.
- Personal hygiene items including toothbrush, toothpaste, comb, brush, soap, contact lens supplies, and feminine hygiene supplies.
- Extra eyeglasses, walking aids, and similar items.

Include anything else that you feel is important. Ask for suggestions when you discuss the plan with your family. Your healthcare knowledge can be a guide. The American Red Cross and other organizations have websites that can provide additional suggestions.

Reassurance for Children

While disasters may cause apprehension for adults, they often affect children more seriously. Children may grow uneasy when they do not know what to expect. They may be especially upset if they have to leave their home. In an emergency, children are often afraid that someone will be injured or killed, that they will be separated from their family, or that they will be left alone.

To help children cope with disaster:

- Comfort and reassure them that they are safe.
- Explain the situation to children. They need to know that they are not responsible for what happened.
- Explain what will happen next. For example, "Tonight, we will all stay together in the shelter." If possible, include the entire family in the discussion.
- Encourage children to talk about their fears.
- Have at least one adult stay with the children before and during an emergency.
- If possible, keep children busy with their daily routines.
- Adults should not show or tell their fears to children.

Summary

Government, institutions, and individuals need to be prepared for emergencies. Individuals must be prepared both on the job and at home. Healthcare workers must be familiar with the emergency plans of their employers and their role in the plans. In addition to the contributions they make to disaster efforts at work, healthcare workers can also be invaluable as volunteers.

Disaster work is stressful. Learning to recognize the signs of stress and how to cope with stress is critical for healthcare professionals both at work and as volunteers. It is possible to reduce one area of stress—concern for family. By being prepared at home and by planning for emergencies, healthcare workers can do their jobs without worrying about their families.

Key Terms

burnout

Community Emergency Response Team (CERT)

credentialed

crisis intervention

critical incident stress debriefing (CISD)

defusing

emergency drills

Emergency Management Assistance Compact (EMAC)

Emergency System for the Advance Registration of Volunteer Health Professionals (ESAR-VHP)

first aid

go bag

just-in-time training

Medical Reserve Corps (MRC)

sheltering-in-place

Reinforcing Terminology

Choose the word or phrase from the list of key terms that best matches each of the following descriptions.

1. A type of certification offered by the American Red Cross ________
2. A tactic used when not enough skilled workers are available ________
3. Nonmedical volunteer opportunity ________
4. Procedure that teaches employees how to put an emergency plan into action ________
5. A mutual aid agreement among states for disaster relief ________
6. Stress-reducing technique lasting from two to four hours ________
7. A requirement for working as a volunteer in a disaster area ________
8. Remaining in a facility during a disaster ________
9. Small collection of essentials for an emergency evacuation ________
10. A condition in which people experience emotional exhaustion ________

Reviewing Concepts

Choose the response that best answers the question or completes the sentence.

1. Which of the following is not a basic medical skill?
   a. taking vital signs
   b. providing CPR
   c. drawing blood
   d. providing ventilator support
2. During a disaster response, directing victims to the appropriate area for treatment would be considered a
   a. professional skill.
   b. basic medical skill.
   c. support skill.
   d. technological skill.

3. Which federal agency requires companies to have a written emergency action plan?
   a. Occupational Safety and Health Administration
   b. Environmental Protection Agency
   c. Federal Emergency Management Agency
   d. National Disaster Medical System

4. Why is preregistration important for assisting in disaster response?
   a. It increases the likelihood that you will be paid on time.
   b. It lets family and friends know your whereabouts.
   c. It helps to ensure that you will know what you will be expected to do.
   d. It tells authorities how many volunteers will be needed.

5. Before traveling to an affected region, volunteers should
   a. make sure that their automobile insurance covers rental vehicles.
   b. get a rabies shot.
   c. create a family emergency plan.
   d. complete their income tax forms.

6. Health professionals who want to preregister to work in a disaster area should contact
   a. CISD.
   b. local or state health department.
   c. EMAC.
   d. OSHA.

7. "Survivor guilt" is an example of
   a. a cognitive reaction to stress.
   b. an emotional reaction to stress.
   c. a social reaction to stress.
   d. a physical reaction to stress.

8. A brief, stress-reducing technique that takes place immediately following a disturbing incident is
   a. defusing.
   b. critical incident stress debriefing.
   c. crisis intervention.
   d. posttraumatic counseling.

9. A go bag should contain
   a. iodine tablets.
   b. a manual can opener.
   c. a warm blanket.
   d. photo IDs.
10. When preparing for an emergency at your home, how many days of supplies should you gather?
   a. 1
   b. 3
   c. 5
   d. 6

Exploring Emergency Preparedness Issues

1. Contact your local or state department of health and find out how you can register to volunteer to help in an emergency. Summarize the necessary steps in one or two paragraphs.
2. Prepare an emergency plan for your family. Include every detail described in this chapter: care for your children and pets, where to meet, an out-of-state contact, and so on.
3. Contact your children's schools and ask for a copy of their emergency plans. Write a short summary of the plans. Then discuss if you think the plans meet the criteria discussed in this book.
4. Go to the disaster readiness and response home page of the U.S. Substance Abuse and Mental Health Services Administration at *www.samhsa.gov/Matrix/matrix_disaster.aspx*. Click on "Tips for Talking about Disasters" and "Risk Communication for Public Officials." Describe in a one-page report what you learned about children, adults, and emergency responders.

CHAPTER 

Emergency Response Procedures

What you will learn:

- What triage is and how it is conducted
- The setup and staffing of a Point of Distribution site
- How to control contamination by applying precautions, using personal protective equipment, and isolating patients
- How the Laboratory Response Network investigates suspicious specimens
- Procedures for evacuating a healthcare facility
- The vulnerability of medical records and ways to protect them
- The ways facilities cope with challenges such as power outages
- How to communicate effectively with patients in the aftermath of a disaster

Emergency response in healthcare involves a variety of special procedures, both within the medical facility and at an emergency site. For example, procedures exist for decontaminating patients, evacuating a medical facility, investigating suspicious specimens in the lab, and more. This chapter introduces you to special procedures that are common in emergency response.

Triage

When many people have been injured at one time, healthcare providers need to rapidly identify which people are most likely to survive, which need immediate treatment, and which can benefit the most from treatment. The procedure for sorting victims according to their condition and the available resources is called **triage**.

People who do not understand the principles and goals of triage can find it difficult both ethically and emotionally to participate. While the goal of everyday healthcare is to provide the best possible care to every patient, in a mass casualty incident, the goal is to provide the best possible care to a group of patients. This is a different paradigm of healthcare. Learning about triage will help you be more comfortable and effective if you are called on to help.

Emergency medical technicians (EMTs) are the allied health professionals best qualified and most likely to participate in triage in most situations. Sometimes, hospital-based teams under the direction of an emergency medicine physician conduct triage. In a disaster, triage can take place in many settings. For example, if a building collapses, triage would begin at the site. A second triage at the hospital may be necessary because the condition of patients may have changed from the time they were initially evaluated and because other patients who are able to leave the scene by themselves will go directly to the hospital.

Triage can also happen in other medical settings. For example, on September 11, 2001, thousands of people, caught in the dust and debris from the collapse of the World Trade Center, sought medical treatment in hospitals and offices of private doctors throughout the New York metropolitan area. Triage occurred in almost every setting when they arrived.

Triage Procedures

Triage is performed by teams led by an emergency physician, EMT, or other health provider. Other allied health professionals

Abbreviations

CDC	Centers for Disease Control and Prevention
EMT	emergency medical technician
FBI	Federal Bureau of Investigation
LRN	Laboratory Response Network
MOU	memorandum of understanding
OSHA	Occupational Safety and Health Administration
POD	Point of Distribution
PPE	personal protective equipment
SARS	severe acute respiratory syndrome
START	Simple Triage and Rapid Treatment
SNS	Strategic National Stockpile

may be assigned to a team to provide support functions. The goals of a triage team are to:

- Establish an area to perform triage.
- Estimate how many patients need to be evaluated.
- Estimate how many ambulances are needed.
- Tag patients.
- Transport patients to the ambulance loading area and then to hospitals.

When there is a mass casualty incident, experience has shown that about one-third of the patients will be critically injured and two-thirds will be noncritical. However, when a building collapses or explodes, the proportions change, and many more people could be critically injured. It usually takes no longer than one minute to triage each patient—sometimes less. This means two people on triage can handle 20 patients in ten minutes.

Several well-established methods for triaging patients in the field exist. Different communities may use different systems for different types of emergencies. For example, for events that involve a suspected contagious disease, triage will involve sorting patients into the following three categories:

- Individuals at risk for infection.
- Individuals already infected.
- Individuals who do not need medical care because they have been immunized, recovered, or died.

In a mass casualty incident involving trauma, such as an explosion, an effective method of triage is **Simple Triage and Rapid Treatment (START)**. Using START makes it possible for only a few rescuers, even those with limited training, to triage rapidly a large number of patients and move them to treatment centers for more detailed assessment. The START system separates mass casualty patients into four clinical categories, each with its own color code. Emergency responders follow clinical guidelines to evaluate each patient. They assess such factors as:

- Ability to walk away from the scene.
- Presence or absence of spontaneous breathing.
- Respiratory rate greater or less than 30 per minute.
- Pulse.
- Mental status assessed by the ability of the patient to obey commands.

A triage tag, which contains the clinical information, is attached to the patient. Rescuers who take over after triage take action based on the color and information on the triage tag.

Triaging patients in emergencies that involve chemicals or other hazardous substances requires special procedures. Contaminated patients—those who have a chemical or other hazardous substance on their clothes and body—should be evaluated for their medical condition before the dangerous substance is removed.

Contaminated patients are triaged in three locations:

- Hot zone—the area where chemical incident occurred.
- Warm zone—an area at least 300 feet from the incident.
- Cold zone—an area next to the warm zone but away from the incident.

A Closer Look

START Triage Categories

Four categories of triage are:

Minor—Green Triage Tag
- Victim with relatively minor injuries.
- Status unlikely to deteriorate.
- May be able to assist in own care.
- "Walking wounded."

Delayed—Yellow Triage Tag
- Transport of the victim can be delayed.
- Victim has serious and potentially life-threatening injuries, but status not expected to deteriorate significantly in the next several hours.

Immediate—Red Triage Tag
- Victim can be helped by immediate intervention and transport.
- Requires medical attention within 60 minutes to survive.
- Patient's airway, breathing, and circulation are compromised.

Expectant—Black Triage Tag
- Victim unlikely to survive given the severity of injuries, the level of available care, or both.
- Victim should be made as comfortable as possible, and pain relief should be provided.

In the hot zone, all healthcare personnel wear personal protective equipment. Very little triage is conducted in the hot zone, except for checking the airway of a patient and trying to stop bleeding. In the warm zone, patients are rapidly sorted into traditional triage categories. Those with the most severe symptoms are treated first. In the cold zone, patients are evaluated for secondary injuries and sent for treatment.

Supporting Triage

Allied health professionals can function in specific roles or supporting roles in triage. For example, respiratory therapists can assess respiratory function during triage. An example of a more general role for health professionals includes helping transport patients to waiting ambulances or nearby treatment areas. This requires equipment such as stretchers. The people who move the victims are known as **litter bearers**; they are formed into teams of three or four people and report to the triage unit leader for their assignments. To transport patients efficiently, an ambulance loading area is established, and people are assigned to keep it organized. Someone will be assigned to coordinate where patients are sent, including communicating with receiving hospitals.

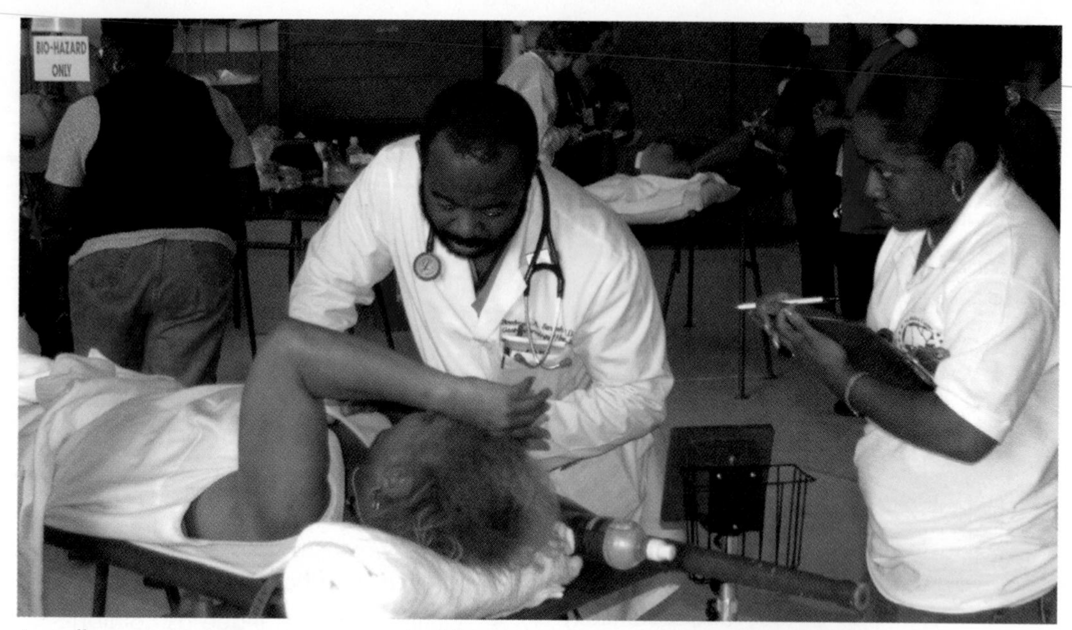

A yellow tag triage section at Dobbins Air Force Base in Georgia shows Veterans Administration doctors, nurses, and other medical personnel evaluating evacuees upon arrival from New Orleans following Hurricane Katrina in 2005.

Working at a Point of Distribution Site

A **Point of Distribution (POD)** site makes the delivery of a special kind of medical care possible. A POD is a temporary station for dispensing vaccines, medicines, or antidotes on a large scale; this is called **mass prophylaxis**. Although a local government agency sets up a POD, it may be supported by state or federal resources. Many health professionals are needed to staff a POD, in both medical staff and support roles.

Strategic National Stockpile

The medications or immunizations disbursed through a POD are likely to come from the federal government through a program called the **Strategic National Stockpile (SNS)**, which is a massive storage of medicines, vaccines, antidotes, and related

During a large-scale training exercise, this Rhode Island Emergency Management Agency Point of Distribution (POD) was set up in Woonsocket, Rhode Island.

medical supplies. Maintained by the Centers for Disease Control and Prevention (CDC), the Strategic National Stockpile has large quantities of **push packs**. Push packs are ready-to-deploy containers with enough medical supplies and medications to treat thousands of patients. Once federal and local authorities agree that the supplies are needed, they can be delivered to any state within 12 hours. Each state has emergency plans for receiving and distributing the medicine and supplies to local communities.

Large packages of drugs, vaccines, medical supplies, and equipment are delivered by air or by ground transportation in two phases. The first phase, the complete package does not include medications and supplies needed for chronic diseases such as diabetes or hypertension. The second phase provides delivery of specific requested medications, sent by the manufacturers under contract with the federal government; these supplies are called **vendor-managed inventory**.

A Closer Look

Roles for Health Professionals in a POD

Some of the roles for health professionals in a POD site are:

- Greeters—Direct people into the POD, provide basic information, and screen for visibly ill people who require immediate medical evaluation.
- Forms Distributors—Hand out and collect patient forms used in triage and follow-up, hand out patient information sheets on medications or immunizations.
- Triage Supporters—Identify people requiring medical or mental health evaluation, identify drug allergies and contraindications for treatments.
- Immunizers—Administer medication.
- Immunizer Assistants—Check supplies, prepare injection site, apply dressing, provide patients with instructions.
- Emergency Medical Technicians—Evaluate and stabilize ill patients, initiate antibiotics and medications prior to transporting patients, respond to medical emergencies.
- Transportation Assistants—Move patients to ambulances, buses, or vans to be transported to a medical facility.
- Medication Dispensers—Repackage medications, determine correct medication types and doses for adults with special medical problems and for children.
- Recovery Area Staff—Help people who have reactions to the medication or vaccine.
- Forms Collection and Exit—Check patient contact information, advise on follow-up care, check that patients received the correct medications.

When a SNS push pack arrives, the state sends teams, including pharmacists and technicians, to the arrival site to inventory and accept the shipment. The team repackages the inventory into smaller packs that are distributed to the PODs or other care sites.

Additionally, some cities and medical facilities have their own stockpiles of medicines.

POD Setup and Staffing

There are two approaches to distribution—"pull" and "push." Most patients are expected to come to the POD if a mass prophylaxis is

A Closer Look

Support Roles in a POD

Some of the support roles in a POD site are:

- Volunteer Coordinator—Coordinate, train, and oversee volunteers.
- Translator—Translate oral and written instructions, help complete patient records.
- Inventory Worker—Maintain store of medications and supplies, including those that require refrigeration; restock dispensing stations.
- Patient Traffic Director—Direct patients.
- Data Entry Worker—Transfer patient information from written forms to computer databases.
- Communications/Information Technology Support Staff—Provide communications links to other PODs and the emergency operations center, support computers used in inventory management and data entry.
- Food Service Provider—Provide meals, snacks, and beverages to staff.
- Security Staff—Control crowds and secure medications, confidential patient information, and communication and computer equipment.

necessary. The federal government calls this the "pull" approach. Health workers can also bring medicine into homes, when people cannot come to the site; this is the "push" approach.

POD locations should be easy to get to and able to accommodate the expected number of patients. A POD site should have entrances and exits clearly marked with security staff located at both locations to maintain order. A POD site should be setup to allow a controlled flow of people. The setup should follow a logical path from entry to the exit.

A POD site requires both medical personnel and large numbers of support workers. Allied health professionals are likely to fill many roles as employees or volunteers. Workers have to be mobilized quickly, so it is important that volunteers preregister with their local health department for a POD team.

As noted in Chapter 2, the federal government may also deploy a National Pharmacy Response Team to supplement local medical professionals.

Controlling Contamination

Contamination occurs when a person, surface, or object has a toxic substance on or in it. The substance may be a chemical or radioactive agent or an infectious organism. Health professionals are trained to protect themselves from the infectious diseases and contaminants that they encounter in their usual work. Emergencies such as chemical spills and emerging infectious diseases can expose them to both familiar and unfamiliar contaminants. Be familiar with protective clothing and procedures, as well as how to decontaminate patients and yourself.

Standard and Universal Precautions

Standard and universal precautions are procedures established by the CDC to prevent the transmission of infectious diseases to healthcare workers and patients. Standard and universal precautions include hygiene such as hand washing, protective clothing and gear, and procedures for handling sharp instruments and needles. All healthcare professionals are trained in standard and universal precautions and can apply these skills in emergency response.

In an emergency, healthcare professionals should assume that all body fluids they encounter are potentially infectious materials. When medical records are not available, healthcare professionals may not know whether a patient has an infectious disease such as HIV or hepatitis A, B, or C. Anyone who provides first aid; cleans up blood and other body fluids; washes contaminated laundry; decontaminates surfaces where human tissue, blood, or body fluids are handled; or handles waste is at risk of contamination.

When providing care at the site of an emergency, responders should be briefed about any contaminants present and how to protect themselves. They should be told what protective equipment to wear, how to identify biohazard waste containers, and what to do if they suspect that they have been contaminated.

Personal Protective Equipment

Personal protective equipment (PPE) is protective clothing and equipment that reduces the exposure of a healthcare professional to hazards. The Occupational Safety and Health Administration (OSHA), which governs workplace safety, requires PPE when other procedures cannot reduce exposure to an acceptable level. PPE includes gowns, goggles, gloves, masks, and special respiratory protection. Most healthcare professionals learn about the proper use of personal protective equipment on the job. During an emergency, PPE becomes even more important.

Without proper PPE, do not rush to help at the scene of an emergency. The appropriate equipment depends on the hazard and the type of work to be done. PPE is usually available in a hospital or clinic, but is not always available at the site of the emergency. After a radiological, biological, or chemical incident, responders should receive special training and equipment before they start working at the scene.

The four levels of PPE are A, B, C, and D. Each level is designed to protect against different types of situations. The incident commander at the site of the incident decides what level of PPE is needed after an assessment of the hazard. For example, if the emergency involves a high concentration of a dangerous substance in the air, a respirator will be selected to provide the best respiratory protection. If there is concern about biological

A Closer Look

Hand Washing without Water

All healthcare professionals know the importance of washing their hands often and correctly to stop the spread of infection. This simple task becomes even more important in emergencies that involve the possibility of contagion or contaminants. However, after some emergencies like a hurricane or flood, water supplies may be temporarily cut off or contaminated. If soap and water are not available, alcohol-based hand sanitizers can help reduce the bacteria on the skin. Note that alcohol-based hand sanitizers kill bacteria, but do not remove them.

A Closer Look

Reporting Human Remains

Healthcare professionals may encounter a body when they are providing care outside a medical facility. Victims of natural disasters usually die from trauma and are not likely to have an infection. However, because of the possibility of infectious disease, bodies should be handled only by workers who have special training and equipment, including a powered air-purifying respirator. If you are on site and encounter a body, notify someone from public safety. Trained workers will place the body in a body bag, place an identifying tag on it, and release it to the medical examiner. Handle human remains with respect and dignity. The goal is to identify the victims, preserve evidence, and reunite families with the remains.

terrorism, protective clothing, including gloves and booties, may also be required.

Level A provides the greatest level of protection and includes a protective suit that covers every part of the body. You have probably seen pictures of workers in "moon suits" responding to hazardous material incidents. These workers were wearing Level A PPE. Level D is the uniform that you wear at work every day. It provides basic protection and includes such things as gloves, boots or shoes, safety glasses, and a surgical face mask.

Protective Clothing When working with patients, wear appropriate protective clothing such as gowns, aprons, smocks, lab coats, or clinic jackets. The appropriate type of clothing depends upon the task and degree of potential exposure.

If patient care is likely to expose you to a splash or spray of blood or other body fluids, wear a clean, nonsterile gown to protect your skin and prevent your clothing from being soiled. If possible, remove a soiled gown promptly and then wash your hands to avoid transfer of microorganisms. A mask and gloves may be indicated.

Firefighters and police are having their eyes treated in a temporary medical center in Manhattan following the World Trade Center terrorist attacks that occurred on September 11, 2001.

Eye and Face Protection Wear face shields and goggles whenever there is potential for flying particles, molten metal, acids or other caustic liquids, or gases. Wearing gear that protects the eyes and face is essential in the aftermath of explosions, during search and rescue, and at fires.

Emergency eyewash facilities should be provided in all areas where eyes may be exposed to potentially dangerous materials, such as in a fire, explosion, or chemical release.

Hand Protection Protect your hands with the appropriate gloves if you are in a situation in which you may be exposed to harmful substances, extreme temperatures that could cause burns, or sharp objects that could cause cuts, abrasions, or punctures. Wear gloves whenever you come in contact with blood or skin that is not intact and when you are handling or touching contaminated items or surfaces.

The right type of glove depends on the type of hazard and the tasks to be performed. For example, if you are cleaning up after a flood, use heavy-duty waterproof gloves. If you are unsure of which kind of gloves to use, check with the supervisor. Wear a nonlatex glove if you are allergic to latex.

If possible, always change gloves after you come in contact with material that may contain a high concentration of microorganisms. Remove gloves promptly after using them, before touching uncontaminated items, and before caring for another patient.

Medical Masks Medical masks protect against infectious organisms and other harmful substances in the air. If an infectious disease epidemic is developing or has developed, healthcare professionals may be instructed to wear medical masks so that they are protected and to reduce the spread of the disease.

Masks also remind people not to touch their faces. Viruses that are not typically airborne can be on surfaces, such as skin, and can be transferred to fingers. Change masks when they become moist. Do not leave a mask dangling around your neck.

If a mask is not available at an emergency site, cover your nose and mouth with several layers of a cotton T-shirt or towel, or use several layers of tissue or paper towels.

A regular medical mask does not provide protection from viruses. Viruses are too small to be caught by the fibers of a mask. Wearing a mask may slightly reduce the chance of a person with a viral infection infecting others because the mask may catch droplets of fluid that are expelled when the person sneezes or coughs. However, an uninfected person who wears a mask is not necessarily less likely to catch a viral disease than someone who is not wearing a mask. To be protected against viruses, healthcare professionals need to wear an N95 respirator mask.

In 2003, hospital staff learned the importance of using the correct masks. It only took a few hours for nine Canadian healthcare workers to contract severe acute respiratory syndrome (SARS) after caring for an infected patient. Six were exposed because they were in the room when a tube was placed into the windpipe of the patient to help him breathe. The droplets forced from the patient's mouth contained a high concentration of the

SARS virus. An investigation revealed that the healthcare workers were wearing masks. However, the masks had not been fit-tested to provide the protection required when treating a patient with SARS.

Respirators A **respirator** is a device designed to prevent the inhalation of very small particles, such as viruses, harmful dust, fumes, vapors, or gases.

Respiratory protection is required at a disaster site when airborne contaminants exceed a certain level established by OSHA. Respirators are also used in healthcare settings to protect healthcare providers from airborne infections. Respirators protect a worker by removing contaminants from the air that is breathed, filtering out chemicals and gases, and supplying clean air from a safe source.

Before using a respirator, a person must be **fit-tested**; that is, fitted and tested for the make, model, style, and size of a respirator. Fit-testing must be done whenever a different respirator is worn and must be redone at least once a year. Workers also have to pass classroom and onsite training before using a respirator.

A volunteer assists with the distribution of masks during cleanup activities following Hurricane Katrina.

Decontamination Procedures

Decontamination is the removal or neutralization of harmful substances such as chemicals agents or radioactive materials. Such substances can penetrate clothing and are absorbed rapidly through the skin. The most effective decontamination is done within the first minute or two after someone is exposed. In most cases, emergency coordinators will tell workers if a dangerous chemical has been released and explain what to do. If healthcare professionals are asked to help with the decontamination of patients, they should be given an initial briefing on PPE, the hazard, and their duties.

Many hospitals depend on the local fire department to handle decontamination. If the event is large, the fire department may not be able to help at the hospital. Therefore, hospitals have to be prepared to contain contaminated patients. They should provide the following:

- Climate controlled shelter.
- Separate areas for men and women so that they can remove contaminated clothing.
- Separate areas for washing off the chemical agent.
- A method for containing and disposing of waste water.

The most important part of decontamination is the timely and effective removal of the contaminant. The first step is for the victims to remove and dispose of their clothing. Clothing and contaminated belongings should be placed in a sealed vapor-resistant bag.

Victims must then shower with plenty of warm water and a hypoallergenic liquid soap. This is followed by more warm water and then a final rinse. Sponges and disposable towels should be used. Patients who cannot shower by themselves will need help. The facility should provide gowns or other temporary clothing.

Despite protective steps, some hazardous substances may still expose health professionals to risk. In 1995, terrorists released the nerve agent sarin on the Tokyo subway. Hospitals were initially misinformed that a gas explosion had caused burns and carbon

A Closer Look

Self-Decontamination

To decontaminate yourself, act quickly and follow the instructions you are given. You will probably be told to:

- Quickly take off all clothing that has any contaminant on it. Cut off any clothing that has to be pulled over your head.
- Place your clothing inside a vapor-resistant plastic bag. Anything that touched the contaminated clothing should also be placed in the bag.
- As quickly as possible, wash any contaminant from your skin, using large amounts of soap and water.
- If your eyes are burning or vision is blurred, rinse your eyes with plain water for 10 to 15 minutes.
- Remove contact lenses and put them with your contaminated clothing. Do not put the contact lenses back in your eyes. Eyeglasses can be washed with soap and water and put back on.
- Avoid contact with people who may have been exposed to contaminants, but who have not yet changed their clothes or washed.
- Move away from the area where the chemical was released until emergency coordinators say it is okay.
- Report the location of the storage bags, but do not handle them yourself.

monoxide. When patients arrived in great numbers to the main receiving hospital, hospital staff members had the patients shower. The patients' clothing was placed in plastic bags and sealed. Staff members used surgical masks and gloves. However, three hours after the release, hospital staff learned that patients had been exposed to sarin. In spite of the protective steps taken, more than one hundred hospital staff members (23 percent) experienced acute poisoning symptoms.

Isolating Patients

A **contagious** disease can be transmitted from one person to another. **Isolation** is separating a person known to be infected with a contagious disease from healthy people. **Quarantine** is the separation of people who have been exposed to a contagious disease and may be infected but are not yet ill.

Isolation is part of daily practice in all healthcare settings. For example, someone with tuberculosis will be put in a separate room from other patients. In emergency response, isolation procedures can be implemented on a broader or more urgent scale in the case of an emerging infectious disease.

Health facilities have standard operating procedures for managing infectious patients called protocols. A facility will have one set of protocols for symptoms such as fever and rash, another for fever and respiratory symptoms, and so on. In an emergency situation, the state health department may order a broader implementation of isolation and quarantine.

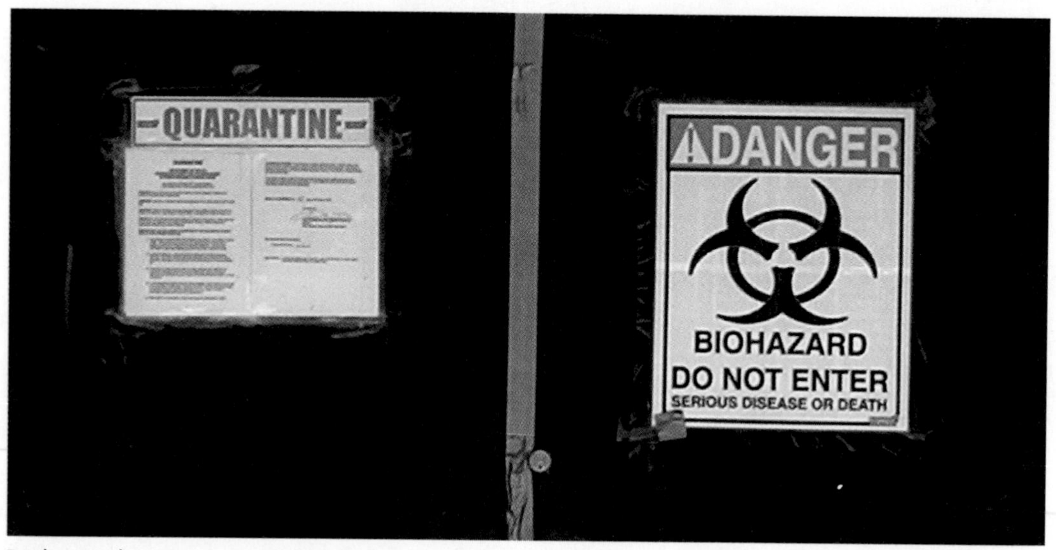

Biohazard signs are posted on the front door of the American Media building in Boca Raton, Florida. Teams of FBI agents and scientists entered the quarantined building that was the former headquarters of *The National Enquirer*. The teams scoured the building for clues about the anthrax attacks that occurred in 2001.

Patients are usually willing to be isolated as long as their personal needs are met; these include housing, food, and the care of their families. However, if patients are a threat to public health and are unwilling to be isolated or quarantined, they may legally be forced into it.

It is important that patients who may have a contagious disease be identified as soon as possible after they arrive at a hospital emergency department. Most healthcare facilities screen patients upon arrival for fever, respiratory symptoms, and rashes. Potentially contagious patients may be given a mask or moved to a separate room.

To protect staff members and to prevent the further spread of disease, it is critical that healthcare professionals consistently use the proper PPE when caring for isolated or quarantined patients. Some facilities require that healthcare professionals observe **droplet precautions**. These precautions mean that they must wear a medical mask when examining a patient with symptoms of a respiratory infection, particularly if fever is present.

Healthcare professionals should always wash their hands before and after all patient contact and should also wash their hands after they come in contact with items that may be contaminated with respiratory secretions. Healthcare facilities should provide tissues and "no-touch" wastebaskets in waiting areas for patients and visitors to dispose of used tissues.

Isolated or quarantined patients should be housed in rooms with separate ventilation systems. **Cohorted patients**, those with similar symptoms, can be grouped together. If possible, privacy barriers such as room dividers should be used to separate patients.

If possible, equipment such as blood pressure cuffs and stethoscopes should be left in the infected patient's room. This equipment should not be used on other patients unless it has been carefully cleaned and disinfected. Healthcare professionals who handle patient linen or laundry should use PPE when they are working in isolation or quarantine areas.

Depending on the type of infection, some patients must be isolated and monitored at a hospital, while others can be treated at home. In the event of a pandemic, there probably will not be enough hospital space to care for all patients. Patients without severe respiratory distress may be isolated at home. Communities have established the criteria for home isolation and quarantine in their pandemic preparedness plans. For example, when patients are at home, their condition must be monitored. Patients or their families may be asked to report their condition by phone or via the Internet to a hospital, clinic, or medical provider's office. Health professionals may be assigned to contact patients for reports.

Medical Laboratory Response

During some emergencies, medical laboratories will handle a much greater number of samples than usual. They must also be prepared to handle unusual specimens, some of which might be hazardous. Medical laboratory technologists may help conduct testing and reporting of suspicious substances.

The importance of laboratories in emergency response was seen in 2001 during an anthrax emergency. Anthrax was sent through the mail, exposing thousands of people to the potentially deadly bacteria. For four months, laboratories were crucial in responding to this crisis. Many agencies depended on the laboratories to provide information that would guide medical, public health, and public safety actions. The laboratories received numerous daily requests for information about the samples they were testing.

Laboratory Response Network

The **Laboratory Response Network (LRN)** organizes public and military laboratories into a three-level system for processing specimens in an emergency. Each level conducts more sophisticated testing than the previous level. The LRN responds to emergencies that involve natural disasters and to events that may involve biological or chemical terrorism. All laboratories in the system use the same protocols and reagents and process specimens in the same way.

The first-level laboratories, known as **sentinel laboratories**, are the hospital, clinic, or private labs where most laboratory

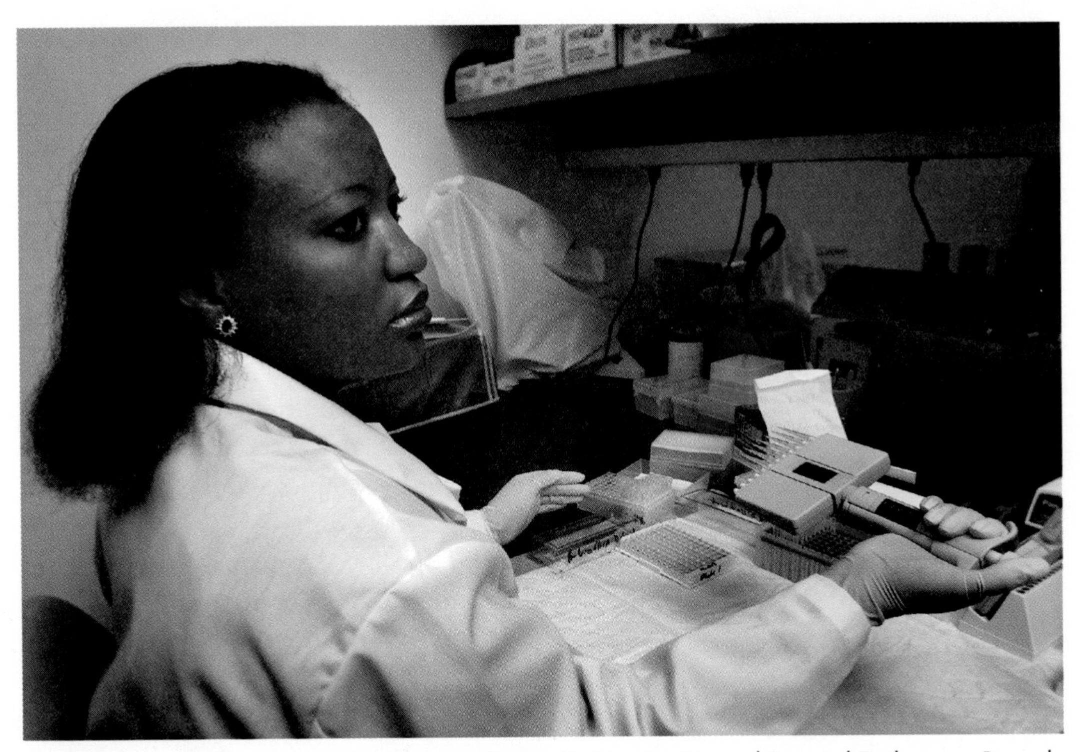

A researcher for the Centers for Disease Control's Meningitis and Special Pathogens Branch, National Center for Infectious Diseases, performs biochemical analyses on samples in an attempt to detect bacterial antigens in 2005.

technicians work and where most medical testing is conducted. About 25,000 routine and diagnostic medical tests are performed daily across the United States. A sentinel laboratory may be the first laboratory to detect a possible biological or chemical agent. It then alerts public health and law enforcement authorities and refers suspicious specimens to the next level of laboratory, which is the reference laboratory.

Reference laboratories have the reagents and technology necessary to conduct more sophisticated tests for biological and chemical agents. Reference labs include state and local public health, veterinary, and military laboratories as well as food, water, and environmental testing labs. If a reference lab confirms that a specimen is a biological or chemical agent associated with terrorism, it follows established procedures and contacts various agencies.

The highest level of laboratories, **national laboratories** are federal laboratories responsible for isolating and identifying specialized strains, such as the Ebola virus, Marburg virus, or smallpox virus. Examples of national laboratories include labs operated by the CDC, U.S. Army Medical Research Institute for Infectious Diseases, and Naval Medical Research Center.

Procedures for Suspicious Specimens

When a suspicious material is collected, it is often brought to a medical lab with its identity being unknown. In addition to conducting some tests, that medical lab is responsible for properly collecting and shipping clinical samples, and for handling specimens that may be evidence in a crime.

Guidelines for handling specific suspicious chemical or biological samples can be found on the website of the American Society for Microbiology. Every lab has procedures for processing suspicious specimens that reflect the following guidelines from the Laboratory Response Network.

- A sentinel lab should never accept a sample that is suspected of being a bioterrorist agent. Such specimens should be sent directly to the nearest reference laboratory.
- If a lab test is being conducted and a possible bioterrorism agent is grown in the lab or detected in any other way, the lab staff should immediately notify the people listed in the emergency lab plan. These may include the treating physician, lab supervisor, manager, director, and infection control officer as well as the state and local health departments.
- A supervisor at a sentinel lab may need to contact the reference lab for guidance in handling a suspicious agent before sending it on for confirmation.

A positive result in a test is considered **presumptive**, meaning that confirmatory tests are needed. In such cases the sentinel lab sends the substance to a reference laboratory. National and international regulations require that infectious materials be properly packaged and labeled before they are shipped. The current guidelines for packaging and shipping are found at *www.asm.org/*.

Many researchers work for the Centers for Disease Control (CDC) in Atlanta, Georgia. The CDC is one of the national laboratories responsible for isolating and identifying the most dangerous pathogens.

In the process of doing their work, labs must coordinate with many agencies. To be effective, labs work with poison control centers, emergency medical service providers, medical toxicologists, departments of health, law enforcement, and emergency management.

A positive result from a confirmatory test sets off a series of actions that include a criminal investigation to uncover the person or persons responsible for the release of the terrorist agent and a public health investigation to determine how many people were exposed to the specimen. Additionally, the best course of treatment for the patient or patients must be chosen.

When a reference or national lab confirms that a specimen is a biological or chemical agent, it notifies the Federal Bureau of Investigation (FBI), state emergency management authorities, and state or local public health officials. The FBI and public health authorities contact the CDC. All this can happen within hours.

Evacuating a Medical Facility

Approximately 20 medical facilities are evacuated in the United States every year. **Evacuation** is the organized, phased, and supervised withdrawal of people from a dangerous or potentially dangerous area. Evacuations of medical facilities may be necessary because of an earthquake, fire, flood, hurricane, release of hazardous materials in the community, utility failures, and so on. Employees are trained in evacuation procedures, and regular drills are held. Learn the evacuation procedures for your facility so that you can help keep yourself and others safe.

Evacuation Plans

Emergency action plans for hospitals, clinics, and long-term care facilities contain evacuation plans. Hospital administrators decide whether or not to evacuate. Evacuating a health facility is challenging and potentially dangerous, so the risks must be carefully weighed. If being outside may expose people to more danger, authorities may decide that it is safer to shelter in place or stay in the building if local roads are not passable. The alternative to an evacuation is sheltering-in-place, which is staying in the most secure area of the facility.

The facility's emergency action plan will include accounting for all staff members and patients after an evacuation. The accounting of all people is critical for two reasons. First, if someone is still in the building, this information avoids delays in rescue. Second, knowing that everyone is out can avoid unnecessary and dangerous search-and-rescue operations.

Evacuations are labor intensive. Health professionals will help notify families and prepare and move patients. Every facility should have a system that notifies key staff members who are not on duty, as well as other staff members if needed.

Some evacuations involve moving people from one part of a hospital to another. Other evacuations involve everyone in the facility—patients, visitors, staff—who will be moved out of the building and transported to a new location. Healthcare professionals will help maintain the continuity of patient care

A Closer Look

Knowing the Evacuation Plan

An evacuation plan contains the following information:

- Actions to take in the event of a fire alarm or "evacuate immediately" order.
- Nearest exits and an alternative route.
- The person in your department who is responsible for taking attendance.
- Your responsibilities for moving patients.
- Special responsibilities such as shutting down critical operations or protecting records or equipment.
- The location of assembly areas inside and outside the building.
- How to check in after a community evacuation. Many employers ask employees to check in through the Internet and to provide contact information.

during an evacuation. They may also help to account for and transport medications, patient records, supplies, and the personal items of patients. They may also have to move equipment with patients, and they may need to improvise if the usual equipment is unavailable. Health professionals may be responsible for turning off or securing medical gases, equipment, medications, lab specimens, or medical records. Such responsibilities should be listed in the evacuation plan.

In an evacuation, healthcare professionals have to consider:

- The condition and physical limitations of the patient.
- Whether a patient needs a ventilator or respiratory support during transport.
- The capacity of the receiving facilities to take care of the sickest patients.
- The distance to the receiving facility.
- Resources available for transport.

Evacuation plans will include a procedure to move patients from their room to an area where they wait to be transported, called a **staging area**, and from the staging area to a vehicle.

If a patient is normally in a wheelchair, it is important to transport the patient together with the chair if possible. Many people who need wheelchairs for daily living have their chairs customized. If a wheelchair is powered, remove the batteries before transporting it and move the batteries with the wheelchair. Be sure that the foot rests are locked and the motor is off. If the wheelchair is left behind, be sure it does not block the path of people. Ask someone else to bring the wheelchair. Reunite the person with his or her wheelchair as soon as it is safe to do so.

The location of each patient must be tracked. Health professionals will complete a tracking form that includes:

- Medical record number or patient identifier.
- Time the patient left the facility.
- Name of transporting agency.
- Original chart sent with patient (yes or no).
- Meds sent with patient (yes or no).
- Equipment sent with patient (list).
- Family notified of transfer (yes or no). Who?
- Private physician notified of transfer (yes or no). Who?

Patients may be transported from the facility by ambulances, buses, or other vehicles. Facilities often have contracts with transportation providers for potential evacuations. Emergency plans also include **memorandums of understanding (MOUs)** with other facilities, which are agreements to receive patients in case of evacuation.

Moving Patients

If a facility is evacuated, you may be asked to help move patients. Many patients will be able to walk on their own. Those who cannot will need to be evacuated by wheelchair or stretcher. In some emergencies, such as fires and earthquakes, elevators

On September 1, 2005, in the aftermath of Hurricane Katrina, the Missouri Air National Guard evacuated patients from a New Orleans hospital. The patients and hospital staff were transported to a hospital in Alexandria, Louisiana.

cannot be used, and patients have to be carried down stairs. A relay team may be needed if the patient has to be carried down more than three flights. Some medical facilities have "stair chairs" that are designed to move people up or down stairs. A person in a wheelchair can be transferred to a stair chair to be moved.

While all patients who are hospitalized or in a long-term care facility have physical limitations that require attention during evacuation, special attention is required for patients with disabilities such as the hearing or visually impaired. Communicate the evacuation to such patients and lead them out of the building to the staging area by holding the elbow of the visually impaired and writing a note for the hearing impaired.

Protecting Medical Records

Medical records contain crucial information that can guide the diagnosis and treatment of patients. They contain patient histories that include chronic or ongoing conditions, allergies, history of infectious diseases, medications and dosages, and so on. Patient records are stored either on paper or electronically on computers.

Challenges

Imagine hundreds or thousands of people arriving for emergency medical care. Even if their records are on file, the staff may not be available to pull the paper medical records. If the patient has not been at the facility in a few years, the paper record may be in storage off-site. Often the facility will never have treated many of the patients, and have no existing records.

In a disaster, paper medical records may be destroyed. The flooding that resulted from Hurricane Katrina in 2005 destroyed the medical records of many people. Electronic medical records that are created and stored on computer are less vulnerable, but may not be available if power goes out, or if computers are left behind in an evacuation, or destroyed by the disaster.

Solutions

Evacuation plans for medical facilities should include procedures for the protection and transportation of patient medical records.

Concerns about the vulnerability of paper records are one reason that many facilities are converting to electronic records. Some facilities are now scanning paper records and storing them electronically. Electronic records must also be protected through remote backup systems.

The loss of medical records after Hurricane Katrina sparked some innovative solutions. The federal government, with help from public and private groups, established a website where healthcare workers treating evacuees could retrieve prescription histories and other health information. The website was assembled by bringing together existing databases from retail pharmacies and government health programs such as Medicaid.

Louisiana also received grant money to develop a health information exchange. All information for patients in New Orleans and Baton Rouge was entered into this system. Anyone who sees patients in this area can see their medical history and a list of medications. Many communities are starting to develop ways to share medical records.

Some health information professionals also found innovative solutions during Katrina, by backing up records on laptop computers, flash drives, or hand-held devices. One nurse practitioner stored the medical records for 100 homebound, elderly patients on her Palm Pilot™. She received calls from providers from across the United States and was able to download and fax the medical history and lab work for these patients to where they were needed.

Health information professionals continue to work to create new procedures that will keep patient records safer, more secure, and more accessible in future disaster situations.

Meeting Facility Challenges

Some disasters may affect the medical facility itself, creating additional safety problems.

Refrigerated Medical Supplies

Power is sometimes lost in emergencies involving wind such as hurricanes and tornadoes, emergencies involving water such as flooding, and in earthquakes. In the event of power outages, medical facilities have emergency generators for essential equipment, but not for everything.

Of special concern are medicines and vaccines that require refrigeration at a constant temperature. It is important for laboratory and pharmacy personnel to know the types of products that will be damaged by lack of refrigeration, including laboratory reagents, disinfectants and sterilants, skin substitutes and burn products, and lock-flush solutions used in intravenous therapy.

A Closer Look

Vaccine

Some vaccine must be kept refrigerated or frozen at all times. If you are responsible for vaccine during a power outage:

- Do not unplug the power cord of refrigerators or freezers, because the unit will resume operating as soon as power is restored.
- If you do not have a secure place with a reliable power source, do not transfer the vaccine. If alternative storage with reliable power sources and secure access is available, consider transferring the vaccine yourself. If you are uncertain about whether the appropriate temperature will be maintained, leave the vaccine in the refrigerator.
- Do not transfer a supply of vaccine from the refrigerator to the freezer. If the power is restored and the vaccine freezes, it may not be usable.
- Do not open freezers and refrigerators until power is restored, unless you are sure that you will transfer the vaccine.
- As soon as power is restored, record the temperature in both the refrigerator and the freezer. Also, record the duration of the outage.
- Do not discard vaccine. Separate questionable vaccine from new vaccine. Store the questionable vaccine in containers that are marked, stapled, and banded until its efficacy can be determined.
- If you are concerned about the efficacy of any vaccine stock, do not administer it until you have consulted with the state or local health department or the manufacturer.

Medical Waste

Power outages and transportation issues can affect the treatment of **medical waste**, which are infectious or biological products that have no further use. Medical waste includes **biohazardous waste**; body fluids or blood products; laboratory specimens from various procedures or biopsies; and **sharps waste** including hypodermic needles, syringes or tubing, scalpels, and pipettes. Medical waste must be handled and disposed of carefully to avoid a potential hazard to public health.

Two primary methods for treating medical waste so that it is no longer hazardous are incineration and autoclaving. **Incineration** is the burning of waste at temperatures ranging from 1,800°F to 2,000°F (982°C to 1,093°C). **Autoclaving** (also known

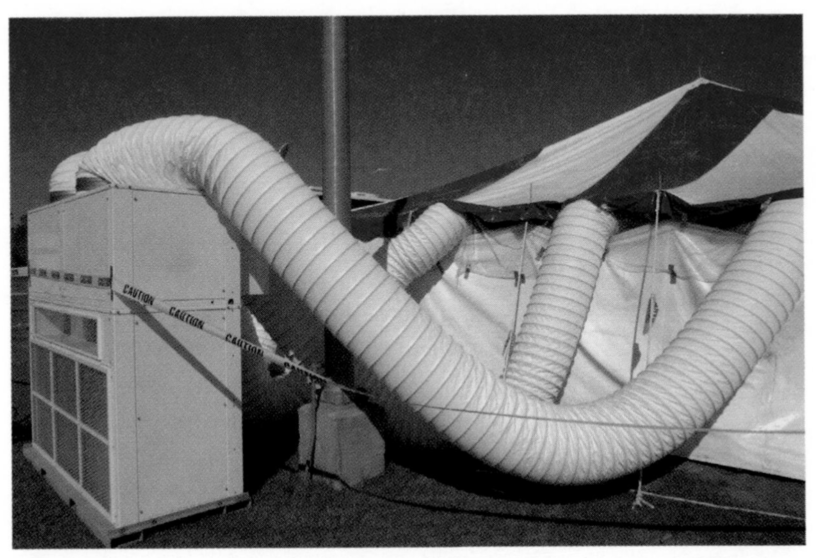

A portable air conditioning unit cools the Disaster Recovery Center in Long Beach, Mississippi on October 22, 2005.

as steam sterilization) exposes the waste to steam at sufficient temperature and pressure and for a long enough time to ensure that microorganisms are destroyed. Medical waste is treated on-site or by outside contractors.

In a widespread emergency, more medical waste is generated than usual because of the increased number of people who need medical care. Facilities or their contractors may not be able to handle the volume. Transportation of waste from the facility may also be disrupted. In its emergency plan, each facility includes alternative procedures for collecting, sorting, transporting, and disposing of waste.

During an emergency, the first priority is the collection and treatment of biohazardous waste because it may become more infectious over time. Biohazardous waste must be treated within seven days. Sharps waste needs to be collected only when containers are full.

Many facilities have autoclaves for sterilizing their own medical waste. However, because these autoclaves are operated by electricity, they may not be working after an emergency in which power is lost. As a result, waste can accumulate before it can be treated or removed.

A Closer Look

Procedures for Reopening

When a medical facility has been damaged, it has to meet a number of requirements before it can reopen. For example, under the Mammography Quality Standards Act, an imaging facility that performs mammograms would have to:

- Inform its accreditation body and the state health department as soon as possible that it is ready to reopen.
- Possibly be reaccredited.
- Document the dates and reasons for lost or damaged records.
- Notify patients whose records have been lost.
- Hire a medical physicist to evaluate equipment that undergoes major repair, is disassembled and reassembled, or is replaced to ensure that equipment is working properly.

In addition to handling their own waste, hospital incinerators sometimes treat the medical waste of private medical practices. In an emergency, if the hospital incinerator is working, it can provide a valuable service to the community by incinerating the medical waste of other medical providers.

If it is not possible to keep up with the treatment of medical waste and power and freezer space are available, healthcare facilities may consider freezing biohazardous waste temporarily.

Communicating with Patients

Your communications skills as a health professional are important during and after a disaster. Here are some procedures for communicating information about emergencies, as well as ways to provide victims with emotional support.

Risk Communication

Risk communication is communicating information about a health crisis to the public in a timely and accurate way that does not unduly heighten concern and fear. Risk communication is

Designated spokespersons usually address the news media. U.S. Department of Homeland Security Secretary Michael Chertoff *(at podium)* is holding a press conference at the Eighth Coast Guard District Offices on Lake Ponchartrain, Louisiana, after reviewing the emergency preparedness plans for Louisiana.

important in many disaster situations, particularly those involving the possibility of contamination or spread of infectious disease.

Individuals, outside agencies, or members of the media may ask you questions about the disaster or the response. Refer all such requests to the designated spokesperson for your facility. The designated spokesperson is usually the public information officer. If you are assigned to answer questions and provide information (for example, through a hotline), you will receive training and written materials to guide you.

Emergencies that involve environmental threats or terrorism are particularly troubling to the public. When there are environmental exposures to hazardous materials, no one knows for sure about the long-term health consequences. When people do not get answers or do not get the answers that they want, they can become angry. Effective communication can help build trust and calm people. Several steps can be taken to communicate effectively in such situations. Most often, a community or facility spokesperson will

A Closer Look

Steps in Risk Communication

1. Be empathic. Show concern for patients and their families.
2. Listen to patients. Try to understand their concerns. Even if patients do not have a perfect understanding of what has happened, you want to understand how the events are affecting their lives.
3. Tell patients that they were right to seek medical care.
4. Provide as much information as possible in a way that the patients will understand.
5. Tell patients about treatment options and help them comply with treatment.
6. Convince patients to cooperate with the treatment.

provide information. However, all healthcare professionals should understand a few basic principles of risk communication in case patients talk to them about events.

Imagine that a chemical agent has had widespread release and that thousands of people need to be decontaminated. Or, imagine that a state agency has ordered mass immunizations or widespread quarantine. State and local emergency management agencies will work with health departments to provide ongoing public education. They might describe what citizens could do and expect to have happen during decontamination, including clothing removal and showers. Often, they will read a prepared statement.

Meanwhile, healthcare personnel will act to reduce anxiety, confusion, and panic among patients. Healthcare professionals may be asked about:

- Current information concerning the exposure.
- Potential short- and long-term effects.
- Recommended treatments.
- Other relevant information.

In speaking with patients, recognize and acknowledge what they are feeling. Before you provide any information, check the facts. Be sure that your source is reliable.

If there is panic or anger, demonstrate assurance and calmness. By keeping cool, health professionals can help patients and others stay calm and work together to solve problems.

Providing Emotional Support

During and after a disaster, patients can become emotionally upset. Some will be experiencing grief over the loss of family members or homes. Grief is a normal response to losing someone or something important. It may appear as sorrow, anger, or confusion. Patients in pain or who have been injured also often have emotional side effects.

Mental health professionals in a community are trained to provide grief services. After an emergency, they mobilize and set up counseling programs and may continue those programs for some time after the emergency.

As a healthcare professional, you may listen to people talk about their disaster experiences and help them resolve their problems. You also need to know when, where, and how to refer patients to other resources. While you cannot offer the type of counseling that is provided by a mental health professional, you can be helpful to patients during your interactions with them.

In the aftermath of an emergency, people often feel better just by talking about the experience. Take in information through what you hear, what you see, and anything else that you pick up or sense. Here are some tips for effective listening:

- Establish eye contact with patients.
- Do not judge what patients are saying.
- Allow silence. Just "being with" the survivors and their experience is supportive.
- Head nodding, caring facial expressions, and occasional "uh-huhs" let patients know that you are listening.
- Paraphrase what you have heard.
- Allow patients to express their emotions, which is an important part of healing. Relax, breathe, and let the survivors know that it is okay to feel what they are feeling.

A Closer Look

Listening Tips

In emergencies or disasters, good listening skills are essential when talking to victims or patients. To show that you are listening when people talk to you about their experiences, you can say:

- "These are normal reactions to a disaster."
- "It is understandable that you feel this way."
- "It wasn't your fault. You did the best you could."
- "Things may never be the same, but they will get better, and you will feel better."

When responding to people, do not sound judgmental or dismiss their feelings. Do not say:

- "It could have been worse."
- "You can always get another pet/car/house."
- "It's best if you just stay busy."
- "I know just how you feel."
- "You need to get on with your life."

Sometimes patients need help from mental health professionals. Speak to your supervisor to find out who might be able to help and what resources are available. Be sure to tell your supervisor about patients you suspect are:

- Disoriented—cannot remember things such as the date or time, do not know where they are, cannot recall events of the past 24 hours or understand what is happening.
- Hearing voices, seeing things, thinking about things that are not real.
- Cannot care for themselves—not eating, bathing, or changing clothes; unable to manage activities of daily living.
- Suicidal or have homicidal thoughts or plans.
- Abusing alcohol or drugs.
- Being abused.

Summary

Health professionals may help support triage, procedures that are used to quickly evaluate a large number of victims. They may also work at a Point of Distribution site in which large numbers of people receive immunizations or medicines. Controlling contamination is always critical, and especially so during an emergency. Healthcare facilities have procedures that must be followed to control and treat contamination, and to isolate or quarantine people. Healthcare facilities must occasionally be evacuated, so employees should know the plans and their job assignments. Also at risk during widespread emergencies are medical records, and healthcare facilities are working on ways to protect these vital records.

Key Terms

autoclaving
biohazardous waste
cohorted patients
contagious
contamination
decontamination
droplet precautions
evacuation
fit-tested
incineration
isolation
Laboratory Response Network (LRN)
litter bearers
mass prophylaxis
medical waste
memorandum of understanding (MOU)
national laboratory
Occupational Safety and Health Administration (OSHA)
personal protective equipment (PPE)
Point of Distribution (POD)
presumptive
push pack
quarantine
reference laboratory
respirator
risk communication
sentinel laboratory
sharps waste

Simple Triage and Rapid Treatment (START)	Strategic National Stockpile (SNS)
staging area	triage
standard and universal precautions	vendor-managed inventory

Reinforcing Terminology

Choose the word or phrase from the list of key terms that best matches each of the following descriptions.

1. body fluids, blood products, laboratory specimens ________
2. hand washing, wearing protective clothing and gear, using good hygiene ________
3. container of drugs, vaccines, medical supplies ________
4. communication information about a health crisis in a timely, accurate way ________
5. wearing a medical mask when examining with respiratory infection ________
6. gloves, goggles, face mask ________
7. used hypodermic needles, syringes, scalpels, pipettes ________
8. Centers for Disease Control and Prevention (CDC), U.S. Army Medical Research Institute for Infectious Diseases, Naval Medical Research Center ________
9. standard operating procedures for managing infectious patients ________
10. withdrawal of people from a dangerous or potentially dangerous area ________
11. massive storage of medicines, vaccines, and antidotes ________
12. agreement to receive patients in case of an evacuation ________
13. method of sorting mass casualty patients ________

14. test result requiring confirmation by a higher-level laboratory ________
15. second phase of federal distribution of medicine during an emergency ________

Reviewing Concepts

Choose the response that best answers the question or completes the sentence.

1. How many patients could a two-person triage team be able to assess in ten minutes?
   a. five
   b. twenty
   c. fifty
   d. two
2. What does the START system use to categorize level of care?
   a. number code
   b. temperature code
   c. color code
   d. alphanumeric code
3. During an emergency, litter bearers
   a. conduct triage.
   b. pick up hazardous debris.
   c. transport victims on stretchers.
   d. drive ambulances to hospitals.
4. Victims of a disaster can go to a point of distribution site to receive
   a. information on the location of family members.
   b. a vaccination.
   c. FEMA-disbursed cash.
   d. a voucher for housing assistance.
5. What is one role of an emergency medical technician at a point of distribution site?
   a. direct people into the POD
   b. hand out forms used in triage
   c. evaluate ill patients
   d. repackage medications

6. Which organization establishes standard and universal precautions?
   a. Centers for Disease Control and Prevention
   b. American Red Cross
   c. Occupational Safety and Health Administration
   d. National Institutes for Health

7. Which item of personal protective equipment requires fit-testing?
   a. moon suit
   b. N95 respirator mask
   c. nonlatex glove
   d. medical mask

8. A patient with tuberculosis should be
   a. isolated.
   b. liberated.
   c. decontaminated.
   d. cohorted.

9. The most specialized laboratories in the Laboratory Response Network are the
   a. sentinel laboratories.
   b. reference laboratories.
   c. national laboratories.
   d. international laboratories.

10. To which laboratory in the Laboratory Response Network should a suspected bioterrorist agent be sent?
    a. sentinel laboratory
    b. reference laboratory
    c. national laboratory
    d. international laboratory

11. During the evacuation of a healthcare facility, the place where patients wait for transport to a safe location is called the
    a. flooding area.
    b. staging area.
    c. commons area.
    d. shelter-in-place area.

12. What is the proper procedure for dealing with refrigerated vaccines during a power outage?
    a. Unplug the power cord.
    b. Transfer the vaccine to the freezer.
    c. Discard the vaccine.
    d. Keep refrigerators shut.

13. A method of cleaning medical waste using steam is
    a. autoclaving.
    b. incineration.
    c. mass prophylaxis.
    d. decontamination.

14. During an emergency, which type of waste should receive first priority in collection and treatment?
    a. sharps waste
    b. biohazardous waste
    c. blood products
    d. medical waste

15. What is the minimum temperature required for incineration of medical waste?
    a. 212°F
    b. 800°F
    c. 1,800°F
    d. 18,000°F

Exploring Emergency Preparedness Issues

1. Contact a local hospital. Find out about their procedures to isolate or quarantine patients. Will these procedures change in the event of pandemic influenza? Does their plan for an episode of pandemic influenza have a specific task that will be performed by your health profession? Discuss what you learn in a one- to two-page report.

2. Contact a medical laboratory in your community. Are they part of the Laboratory Response Network? Have they handled a "suspicious sample?" What were the tasks carried out by your health profession? Where is the reference laboratory in your region? Summarize this information in a one-page report.

3. Contact the local emergency medical services agency in your community. Find out about their procedures for decontaminating victims in the field. How often have they had to use these procedures in the past two years? What equipment is used to communicate to the local hospital? Does your hospital include your health profession in their plan to decontaminate patients? If yes, what are the tasks? Summarize what you learn in a two-page report.

CHAPTER 5

Continuity of Care

What you will learn:

- How to advise patients with chronic diseases to prepare for and cope with disasters
- What special needs shelters are, how they are staffed, and how they operate
- The healthcare needs of other vulnerable groups

When we think about health-related problems and disasters, we usually think about injuries and illnesses that are the immediate and direct result of a disaster. We rarely think about how disasters affect health issues on a broader scale. For example, people who are chronically ill have special health needs that may not be met or met easily during and after a disaster. Health professionals strive to provide patients with ongoing, coordinated care from the same group of providers when needed. This type of care is called **continuity of care.**

Patients with Chronic Conditions

A disaster and its aftermath can create many challenges for people with a **chronic disease** which is a disease that is long lasting and treatable, but not curable. Examples of chronic diseases are heart disease and diabetes. People with chronic diseases are at higher risk of injury, infection, and stress than other people. They depend on medications, medical equipment, and special treatments that may not be available during or after emergencies. Health professionals can help such patients prepare for potential emergencies.

Health Complications

Disasters put patients who are chronically ill at risk. These patients are affected by the emergency itself and by the disruption of their usual medical care. Certain health complications should be expected and planned for. For example, patients with cardiovascular conditions may have higher blood pressure or unstable angina during or immediately after an emergency, thereby increasing the risk of myocardial infarction or stroke. People with asthma and other pulmonary conditions may have problems that are triggered by exposure to mold, smoke, or airborne pathogens. People with diabetes are vulnerable to infection if they get an open cut or wound. HIV, chemotherapy, and some medications lower immunity, which might put those patients at increased risk in the event of an emerging infections disease.

In response to the dangers associated with Hurricane Katrina in 2005, more than 200,000 people with diabetes were evacuated. Continuity of care for these people was impossible, because many lost their regular source of medical care and went to shelters without access to needed insulin.

After a disaster, healthcare facilities should expect a surge in the number of chronically ill patients who arrive for treatment. This is true for hospitals and private medical practices, assisted-living facilities, nursing homes, emergency shelters, hospices, dialysis centers, and other facilities. Patients may need:

Abbreviations

EMS	emergency medical services
EMT	emergency medical technician

- Prescription medications, medical devices, and treatment such as dialysis.
- Home care and monitoring, physical therapy, convalescent care, hospice services.
- Medical supplies such as compression stockings, diabetes testing supplies, or wheelchairs.

Medical Emergency Kit

As part of emergency preparedness, medical professionals should advise patients with chronic illness to pull together a personal medical emergency kit. Each kit should contain properly stored medications, a list of all medications taken and the dosages, insurance information, and medical records obtained from health providers. The kit should also contain medical supplies for a minimum of one week. For example, a person with diabetes should pack alcohol swabs, syringes, cotton balls, testing strips, blood glucose meters, a pen or pencil, and a pad of paper to record blood sugar levels. Patients should be prepared to store refrigerated medications in a cooler. They should always have a supply of frozen ice packs ready to go.

Patients should inspect their kit at least once a year. They should replace expired medications and update insurance information and medical records. Patients should keep copies of paper records such as insurance cards and prescriptions in a waterproof plastic bag to prevent water damage. Patients will also want to keep a list of their healthcare providers and contact information.

The emergency kit should be portable and stored in a place where it can be easily accessed. Family, friends, and neighbors of the patient should know where the emergency kit is kept.

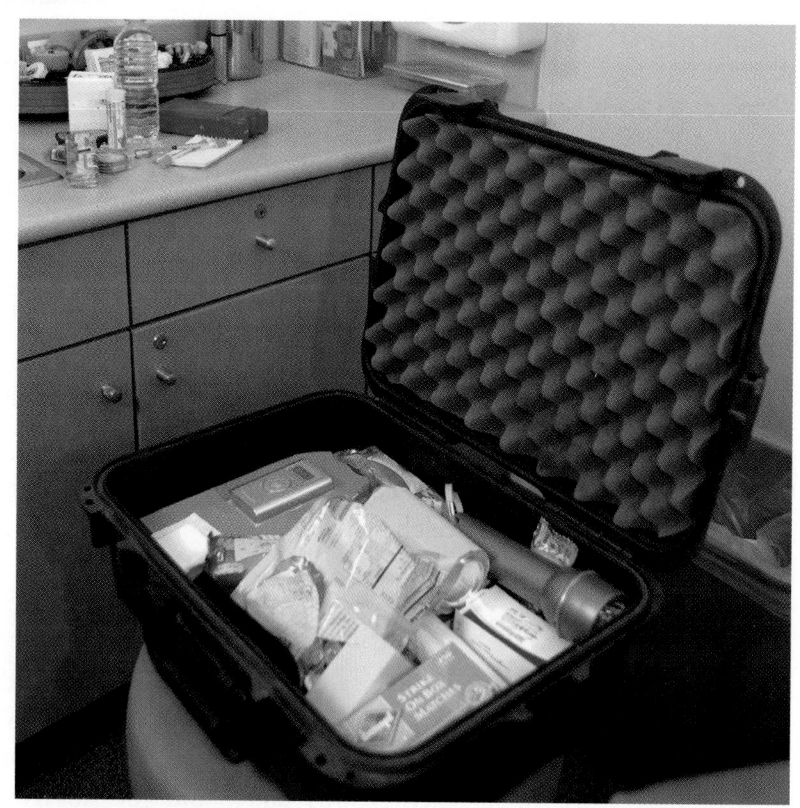

A sample "Diabetes Disaster Kit" shows some recommended items including blood glucose testing strips, copies of prescriptions, and an emergency kit.

Medications and Supplies

Not everyone will be prepared for an emergency. Communities have to be prepared to provide medication when people are without it. Replacing lost medications can be a problem. Prescription insurance plans often limit the number of refills or the amount of medication that can be dispensed in a single month. If a person has already had a prescription refilled, the insurance company may not immediately pay for another refill that month.

Healthcare professionals should remind patients that extremes of temperature and humidity can affect medications and supplies such as blood glucose monitors and test strips. Some medications must also be protected from direct heat and sunlight.

A Closer Look

Insulin

Insulin should be stored between 35 and 46 degrees Fahrenheit. It may be left unrefrigerated (between 59 and 86 degrees Fahrenheit) for up to 28 days and still maintain potency. If power is lost, insulin should be kept as cool as possible, away from direct heat and out of direct sunlight. Ice can be used to keep it cool, but insulin should not be frozen. Questionable vials of insulin should be discarded and replaced when new insulin is available. Patients or healthcare providers who have specific questions about the suitability of their insulin can contact the manufacturer.

If electrical power has been off for a long time, medications that require refrigeration may have to be discarded. Other medications that require refrigeration may still be viable, but they will have a shorter **shelf life**, which is the amount of time a medication is usable before it reaches its expiration date. Insulin, for example, may not have been refrigerated if electrical power has been off, but it may be used until a new supply is available. Package inserts for medications and supplies may provide information about what to do if the medication has not been refrigerated. For this reason, patients should be encouraged to keep package inserts or the original medication containers. These might also prove helpful for people replacing medications without a prescription.

Any medications that come in contact with contaminated water should be discarded. Medical supplies such as blood glucose meters and test strips, sterile gauze and wound dressings, disposable thermometers, syringes, and tubing should not be used if the packaging is not dry and intact. Heat and humidity can also harm medical supplies. For example, diagnostic tests, including blood glucose tests, may not be accurate if the meter is damaged. Meters should be discarded if they have been exposed to unusually high temperatures for more than 24 hours. Patients should consult with a pharmacist or supplier of durable medical devices about whether or not to dispose of a device.

Electronic Equipment

Patients who depend on equipment that operates on electricity, such as ventilators, are particularly vulnerable to power outages. Advise these patients to plan ahead and:

- Notify their electric company and fire department or emergency medical services (EMS) agencies that they have a medical device that needs power.
- Check user instructions or call the distributor or manufacturer to find out whether the device can be used with batteries or a generator.
- Locate and hook up a generator if possible.
- When the power is restored, check to be sure that the settings on the medical device have not changed. Medical devices often reset themselves automatically when power is interrupted, as do digital clocks.
- As area power networks are restored, voltage levels may fluctuate. This may cause devices to continue to reset or operate improperly. Be sure patients check this.

If electronic devices are exposed to water or high levels of humidity, they may not operate properly. Before using such a device, patients should:

- Immediately turn off the power at the main circuit breaker if electrical circuits and equipment have gotten wet.
- Dry the device thoroughly. Wipe the outside of the device with a dry cloth.
- Check devices for visible contamination.
- Check all power cords to be sure they are not wet or damaged by water. If they are damaged, do not plug them into an electrical outlet.
- Check all settings and alarms to be sure they are working properly.
- Run quality control checks often to be sure the device is working properly.

Special Needs Shelters

A **special needs shelter** is an evacuation shelter outfitted with special equipment and medical staff to care for people who cannot be properly cared for in the evacuation shelters set up for the general population. State health departments are responsible for setting up these temporary facilities after a disaster involving evacuations. This kind of shelter serves people who:

- Have medical conditions that require professional observation, assessment, and maintenance.
- Have chronic health conditions that require assistance.
- Require skilled nursing care but not hospitalization.
- Need professional assistance for medications or reading of vital signs.

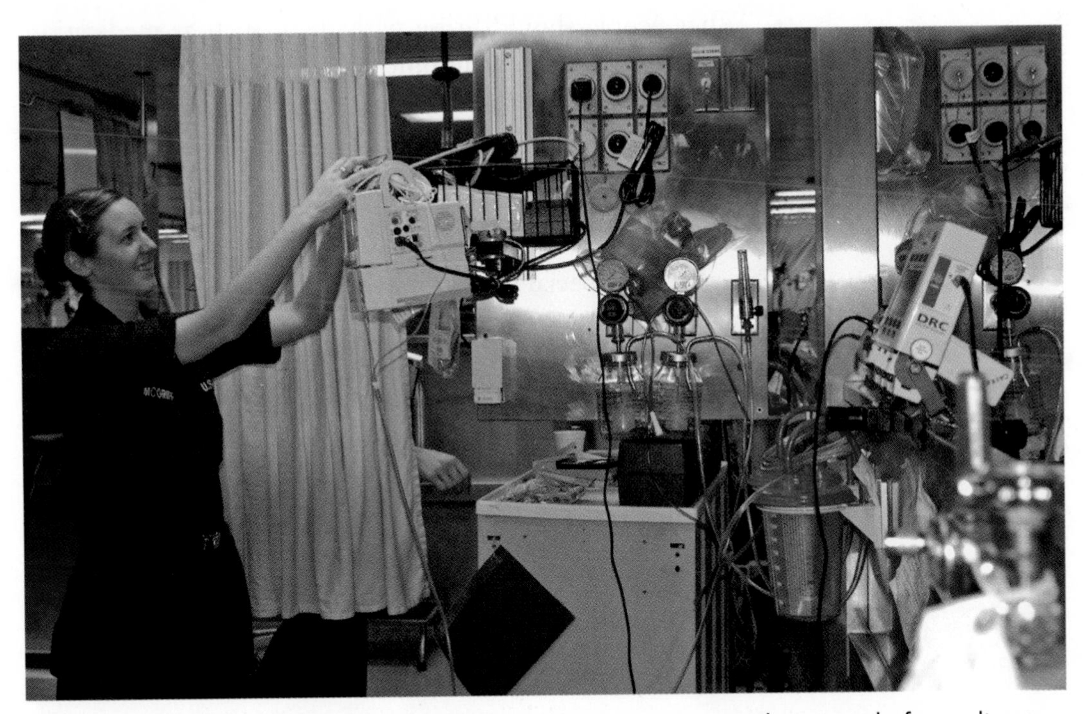

People with chronic respiratory conditions need continuing care during and after a disaster, and their conditions may worsen because of the impacts of the disaster. This oxygen delivery system, on the hospital ship USNS Comfort, was brought to New Orleans after Katrina in order to help provide such care.

The setup, mission, and operation of special needs shelters vary from state to state. Patients at special needs shelters may require oxygen, special diets, disability accommodations, or refrigerated medications. Special needs shelters need a refrigerator and cupboards for the storage of medication and arrangements for oxygen and its storage.

Not all special needs shelters can provide the equipment necessary for dialysis or ventilation. Patients who need such treatment will probably have to be hospitalized or should make arrangements to go to a facility outside the evacuation area.

Operations Plans

The state departments of health have procedures for educating the public about special needs shelters. The public will be told:

- Who is eligible to go to special needs shelters.
- What people should bring with them such as medication, medical supplies and equipment, special dietary foods, water, personal items such as eyeglasses, hygiene articles, change of clothing, etc.
- Whether the caregivers for special needs patients should accompany them.

The plans for special needs shelters will include procedures for the day-to-day running of the shelter and the conditions under which patients should be sent to hospitals. Arrangements may be made for the local emergency medical services to transport residents from the shelter to the hospital.

In some areas of the country, residents may be asked to register ahead of time if they have special medical needs so that they will already be assigned to a shelter when a disaster such as a hurricane hits.

Staffing

Special needs shelters are staffed with a multidisciplinary team of medical personnel. To ensure that these shelters can be adequately staffed, communities may contract in advance for personnel with

A Closer Look

Emergency Generators

Special needs shelters often rely on emergency generators to provide power for medical equipment. Following Hurricane Katrina in 2005, a study showed that emergency generators seldom provided uninterrupted electric power to the shelters. Some failures were the result of lack of maintenance of the equipment before the storm. Most generator failures occurred because the generators had to operate continuously for an extended time.

Well-maintained generators are an important part of emergency preparedness. The following procedures must be followed to prevent carbon monoxide poisoning from generators:

- Never run a generator or any gasoline-powered engine inside an enclosed structure, even if the doors or windows are open, unless the equipment is professionally installed and vented.
- Never run a generator or any gasoline-powered engine outside an open window or door where exhaust can waft into an enclosed area.

needed skills, such as the local EMS agencies. The staff may also include volunteers from the community. Typical staffing may include emergency medical technicians (EMTs), respiratory care personnel, physicians, and nurses. Many support staff workers are also needed. Their duties include creating records, maintaining paperwork, distributing food, and so on.

Patients may be asked to bring someone who will remain with them at the shelter. They may also be asked to bring their own supplies if possible. Their caregivers are expected to assume primary responsibility for personal care tasks, such as dressing, walking, assisting patients to the bathroom, personal hygiene, and feeding. The sheltered person or their caregiver is responsible for administering routine medications, changing dressings, and handling special equipment. Patients would be assisted by healthcare professionals as needed. The sheltered person or the caregiver would be responsible for managing the oxygen and equipment, assisted by respiratory care personnel or EMTs.

Caring for Vulnerable Groups

People with disabilities, the elderly, children, and people who do not speak English have special healthcare needs during and after a disaster. Health professionals should be aware of these special needs during emergency response.

People who have limited mobility or who are hearing or visually impaired may need help to evacuate or to find medical care. They may have difficulty learning about the emergency or what to do, and they may need help with transportation if an evacuation is ordered. For example, after Hurricane Katrina, many people who were hearing-impaired did not know about the evacuation instructions. Because power was out, they could not follow closed-captioned television announcements. Cell phone text messaging also did not work.

A FEMA employee assists displaced children following the Northridge Eathquake, which struck southern California on January 17, 1994.

A Closer Look

Survival Kits for the Elderly

In 2005, when Hurricanes Katrina and Rita devastated the Gulf Coast, the Austin, Texas, area did not have to evacuate, despite extensive flooding. The elderly and people who were homebound were unable to leave their homes for several days because of the flooding. Many people came close to starving because they did not have enough food or emergency supplies. Afterwards, the Aging Services Council of Central Texas created kits so that elderly residents could survive future short-term emergencies such as floods, power outages, thunderstorms, and ice storms. Each kit contained three days of nonperishable food, a flashlight, batteries, toilet paper, and telephone numbers to use in the event of an evacuation. Volunteers assembled 800 kits and distributed them to low-income adults age 60 and over who lived alone.

People who speak limited English are another vulnerable group. Federal regulations require hospitals and medical facilities to provide interpreters and signs in many languages. Organizations that print public information as part of disaster preparedness should include information in languages spoken in their area.

Children who have become lost or separated from their families are at special risk after a disaster. For example, children typically cannot receive medical treatment without parental permission. Facilities usually have protocols for treating children whose parents are not present.

Elderly people may need special attention and require help with meals or transportation. The elderly also are sensitive to extremes of hot and cold weather if heat or air conditioning is disrupted.

Summary

Healthcare professionals need to know what advice to give people with chronic illnesses so that they can be prepared to deal with disasters. Those with chronic medical conditions should prepare medical emergency kits containing their medications and medical supplies. Those who cannot be cared for in general shelters also need to understand that special needs shelters are available and that these are the places they should go in case of an evacuation. People who are not able to care for themselves, such as unattended children, the elderly, or the disabled, have special needs that must be addressed during emergency situations.

Key Terms

chronic disease
continuity of care
shelf life
special needs shelter

Reinforcing Terminology

Choose the word or phrase from the list of key terms that best matches each of the following descriptions.

1. period of time in which a medication is viable ________
2. facilities outfitted with equipment and staff to care for patients who cannot be cared for in general shelters ________
3. type of medical care that is often interrupted during a disaster ________
4. a health condition that is long lasting and treatable but not curable ________

Reviewing Concepts

Choose the response that best answers the question or completes the sentence.

1. An example of a chronic disease is
   a. the common cold.
   b. influenza.
   c. broken bone.
   d. asthma.
2. During an emergency, all of the following would be considered patients with special needs *except* people who need
   a. corrective lenses.
   b. skilled nursing but not hospitalization.
   c. their vital signs read.
   d. professional medical observation.
3. A personal medical emergency kit should contain a
   a. medical ID bracelet.
   b. year's worth of medical supplies.
   c. list of healthcare providers.
   d. list of the medications of family members.
4. Who has responsibility for setting up special needs shelters?
   a. federal government
   b. state health department
   c. American Red Cross
   d. private healthcare contractors
5. An essential piece of equipment for special needs shelters:
   a. defibrillator.
   b. ventilator.
   c. refrigerator.
   d. generator.

Exploring Emergency Preparedness Issues

1. Identify the groups of patients cared for by your health profession who could be defined as "vulnerable" in the event of a disaster. Describe their special needs during a disaster. What would happen to their care if there was no power and/or no safe water or food? What resources would these patients need? Which agencies in the community could be helpful in planning to care for these patients? What plans would need to be developed in advance to reach out to your patients? What services could your profession offer? Summarize your response in a two-page report.

CHAPTER 6

Skills and Roles

In this chapter, you will learn how the skills of different allied health professionals apply to emergency response. The professions covered are:

- Emergency medical technician
- Health information technician
- Laboratory technician
- Medical assistant
- Medical imaging and radiation therapist
- Pharmacy technician
- Physical therapist assistant
- Respiratory therapist
- Surgical technologist

All allied health professionals have skills that can be applied to emergency response. Their roles could vary, depending on the community and the emergency. This chapter provides information on a range of allied health professions and how the members of each health profession can help in times of crisis.

Emergency Medical Technicians

Emergency medical technicians (EMTs) generally have more training and responsibilities in emergency response than other allied health professionals. They provide rescue, assessment, care, and transportation to emergency departments or other care sites, while working under the direction of protocols provided by an emergency physician. EMTs are trained to evaluate the condition of a patient, and maintain airway, breathing, and circulation by applying cardiopulmonary resuscitation (CPR) and defibrillation. They control external bleeding and prevent shock or further injury or disability by immobilizing potential spinal injuries or

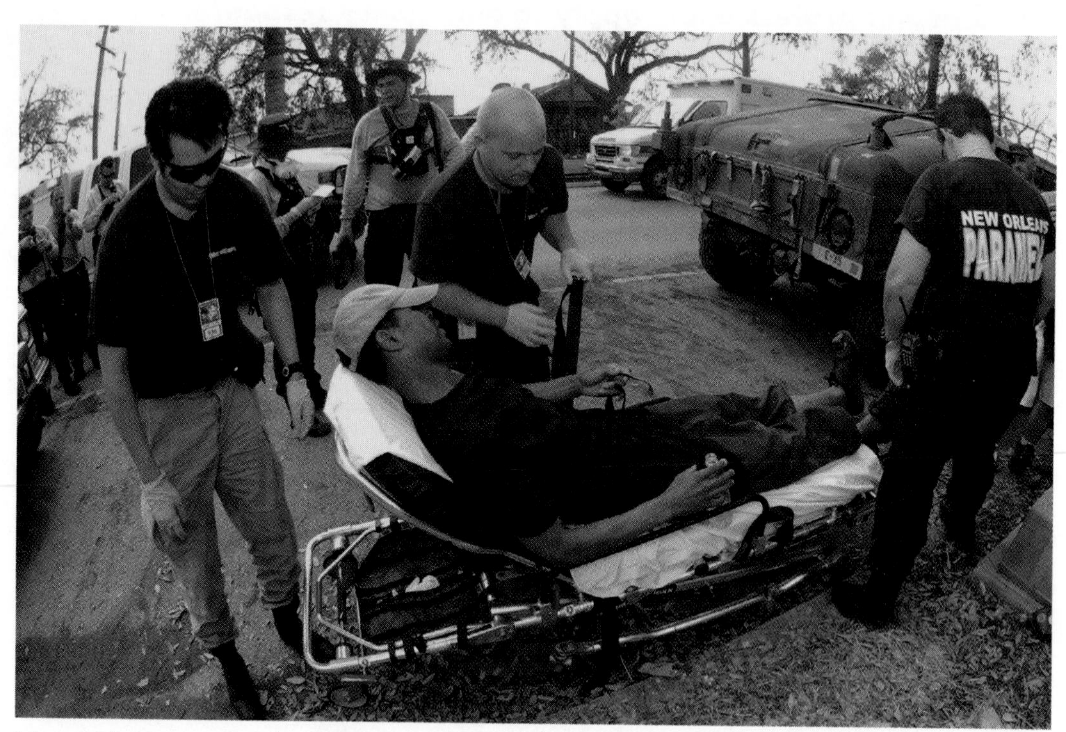

New Orleans paramedics tend to a New Orleans man who was found alive in his home two weeks after Hurricane Katrina struck Louisiana in 2005. Paramedics and other EMTs are often first responders in emergencies.

Abbreviations

ALS	advanced life support
CPR	cardiopulmonary resuscitation
CT	computer tomography
DMORT	Disaster Mortuary Operations Response Team
EMS	emergency medical services
EMT	emergency medical technician
MR	magnetic resonance
POD	Point of Distribution

bone fractures. EMTs also help the staff of hospital emergency departments by obtaining medical histories and administering medical treatment before a patient arrives.

Paramedics are EMTs who have additional training and responsibilities, including emergency advanced life support (ALS) treatment. They may perform many medical procedures at the scene of an emergency or in the ambulance on the way to the hospital. Like EMTs, paramedics get instructions and orders via radio from a doctor at the emergency department or dispatch station. Paramedics can administer some drugs orally or by injection. They can insert breathing-aid devices, use stomach suction equipment, use and interpret heart-monitoring equipment, and apply other emergency medical procedures during the ambulance ride.

Emergency medical technicians and paramedics work in emergency medical services (EMS) and provide care outside a hospital setting and transport patients to the hospital. In an emergency, some EMTs may be asked to help in the emergency departments or intensive care units of hospitals. EMTs may also be asked to provide vaccinations or medications at a Point of Distribution (POD) site if there is a major outbreak of an infectious disease.

Health Information Technicians

Health information technicians create and maintain patient records and collect and maintain medical data. In a disaster, health information professionals preserve existing medical records and create records for new patients.

In disaster response, medical records are often not available to health providers. For example, after Hurricane Katrina in 2005, many health responders did not know the medical histories, allergies, or medication records of patients. This lack of knowledge hampered their ability to care for patients.

Because all medical care has to be documented, in emergencies a large number of new medical records will likely have to be created very quickly. Health information professionals are responsible for maintaining these emergency records, keeping them in a safe place, protecting them so they do not get lost, and making them available, even after the crisis is over.

Additionally, health information technicians can organize and implement the medical records process at a POD site, special needs shelter, or other temporary health site.

Health information technicians also work to protect health records. Paper records are especially vulnerable during many disasters. For example, during Hurricane Katrina and the subsequent flooding, thousands of paper medical records were destroyed. This is one of the many reasons that many hospitals and clinics are working to change over to electronic records. However, access to electronic health records requires electric power. In both hospitals and clinics, backup power, such as a generator, is essential.

To prevent electronic records from being lost, they must be backed up both internally and externally. This means that in addition to backup within the system, all data should be stored up in a location outside the state or region. Servers also should have a built-in slow shut down to avoid data loss in case of a sudden power outage. Implementing such systems is an important role in emergency preparedness.

Laboratory Technicians

Laboratory technicians conduct tests that help detect, diagnose, and treat many diseases. Most work is done in hospital labs. During an emergency, laboratory technicians will have to preserve the work already done, assess which tests they can still perform and how quickly, and prioritize and carry out the work.

During a disaster, medical laboratories may have to process a large number of specimens quickly. If there are many injuries, labs may have to type and cross-match hundreds or thousands of blood samples so that transfusions can be performed. If a chemical accident occurs, a lab might be asked to analyze thousands of samples of body fluids for chemical content.

A laboratory technician conducts tests for bird flu at the University of Pennsylvania, New Bolton Center Poultry Laboratory in June 2006. Lab technicians will provide many important services during some disasters.

Once the samples are processed, laboratory technicians communicate test results to agencies such as local and state departments of health or the Centers for Disease Control and Prevention, move samples to and from the lab, and help with the proper and timely disposal of medical waste. Laboratory technicians should know the protocols for notification and communication of results.

A related professional is the *phlebotomist* who collects blood specimens so that they can be tested. Safe and careful blood specimen collection skills may be much in demand in many disaster situations, especially when infections are involved.

Medical laboratory technicians also have another role in disaster response—identifying suspicious substances. Recall from Chapter 4 that a lab in a hospital or clinic refers to a reference laboratory for advanced testing to confirm an environmental or a biologic agent. Resources are available on the Web to learn about special testing for suspicious or chemical agents. One resource is the **National Laboratory Training Network (NLTN)** at *wwwn.cdc.gov/nltn/*. The National Laboratory Training Network provides laboratory training courses in clinical, environmental, and public health laboratory topics. For information about training resources, call 800-536-NLTN (6586).

Medical Assistants

Medical assistants perform a variety of administrative and clinical duties in a medical office. They schedule and receive patients, maintain medical records, handle telephone calls, take patient histories and vital signs, prepare patients for a variety of procedures, assist the health provider with examinations and treatments, collect specimens, perform some diagnostic tests, and administer medications as directed by the health provider. In some states, they may give injections and draw blood. In other states, some medical assistants may be temporarily authorized to give injections and draw blood by the state department of health. All these skills are applicable to emergency response.

Medical assistants can perform a wide variety of roles during emergencies because of their experience in working with patients.

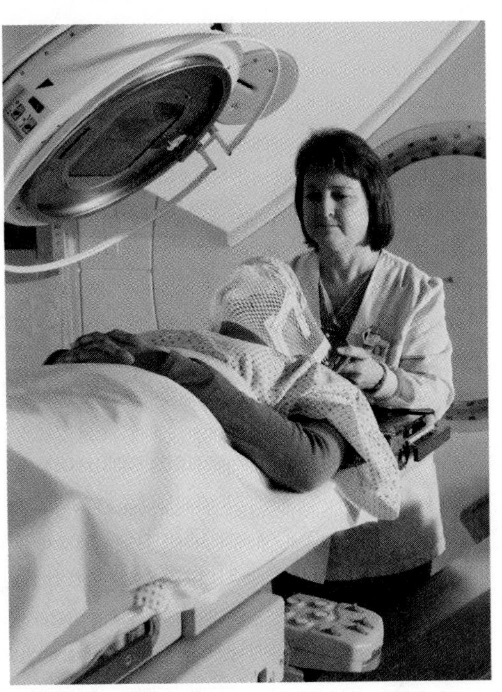

The skills of radiation therapists are in high demand during and after emergencies. In addition to performing a variety of imaging tests on patients, many imaging and radiation therapists are part of Disaster Mortuary Operations Response Teams, which assist in victim identification.

If a disaster occurs, the administrative area of a medical office will be a busy place. Many patients will contact their usual clinic or medical practitioner for treatment, information, or referrals. Medical assistants could communicate with patients about evacuation orders, hazards, precautions, and emergency sources of medical equipment such as oxygen. They may also help patients who need medical care for chronic diseases. Many appointments might have to be changed to accommodate patients needing emergency care.

Medical assistants can apply their clinical and administrative skills as volunteers on medical response teams. They can also perform many nonmedical support roles effectively because of their experience in working with patients.

Medical Imaging and Radiation Therapists

Health professionals in medical imaging and radiation therapy perform diagnostic imaging examinations and administer radiation therapy treatments. These health professionals include *radiographers* and *sonographers,* and they are trained in anatomy, patient positioning, examination techniques, equipment protocols, radiation safety and protection, and basic patient care.

Many types of emergencies create a need for a large number of diagnostic imaging tests to be performed quickly. For example, windstorms and earthquakes, and the clean-up afterwards, are associated with many injuries. Explosions can cause blast injuries that have to be evaluated through imaging tests. Medical imaging also becomes critical if thousands of chest X-rays are needed because of a widespread infectious disease such as influenza. Imaging professionals will respond in their usual work settings or as preregistered volunteers in a variety of locations.

The ability to conduct these tests is dependent on electrical power. Radiology departments were severely impacted during the blackout of 2003, which affected one-seventh of the U.S. population, primarily in the northeast states and adjacent Canada. At Memorial Sloan Kettering Cancer Center in New York City, all patients were immediately removed from the radiology department using lights powered by generator. The radiology department performed emergent procedures by using emergency generators that supported only X-ray machines.

Imaging professionals have helped identify bodies by taking dental X-rays and comparing them with dental records. X-rays and other imaging tests may also produce forensic evidence that is important to police investigations. For example, after the Oklahoma City terrorist bombing on April 19, 1995, 60 radiology technologists took 168 whole body radiographs of those who

A Closer Look

Hurricane Katrina Response

After Hurricane Katrina in 2005, teams of imaging personnel stayed at the Ochsner Medical Center in New Orleans. The teams included two computer tomography (CT) technologists, one magnetic resonance (MR) technologist, and four radiographers. The hospital disaster plan assumed that equipment would be functioning because there was a generator for backup power. However, when the air-conditioning failed, equipment was damaged and the facility was left with only one CT scanner, digital film and ultrasound units, and a backup Windows-based recorder. This equipment was used to conduct 10 CT, 5 ultrasound, and 100 radiographic exams a day for relief workers and residents who had not evacuated.

Radiation technologists were also critical in identifying dead bodies after Hurricane Katrina. Those working on the DMORT had no access to digital radiography. They were able to use only cassette and darkroom processing. This equipment had been in storage for years. There were malfunctions, dead batteries, and no spare parts. The film was outdated, and no one was trained in equipment repair. Despite the difficulties, the teams imaged 15 to 20 bodies per shift with an average of 10 images per victim.

had died. The radiographs helped with both identification and police investigations. Such services are invaluable to the families of victims.

Imaging professionals may volunteer for a Disaster Mortuary Operations Response Team (DMORT) (see Chapter 2). Some radiographers have volunteered to help other health professionals and responders, such as fire fighters, understand radiation and Geiger counters.

Imaging technologists can also apply their basic medical skills in disaster response by checking vital signs, drawing blood, and performing CPR. Such skills are needed in abundance during a disaster.

Radiological Emergency Medical Response Teams are organized to respond to radiological emergencies in many states. To learn more, go to *www.remm.nlm.gov/index.html.*

Pharmacy Technicians

Pharmacy technicians assist pharmacists in filling prescriptions and can help distribute medicines, vaccines, and other agents during a disaster.

As part of preparedness, pharmacy technicians may be asked to help prepare a pharmacy to function without electricity. Most pharmacies use computers to keep patient records and fill prescriptions. Records can be backed up electronically onto a flash drive that can be plugged in at another location. The technician should print out a customer roster every month and store the rosters in a safe location, such as their personal go bag. The roster for each client should include:

- Full name.
- Date of birth.
- Name and dosage of each medication.
- Name and contact information for prescribing physician.
- Date and amount of medication dispensed.

After a disaster, people may not have access to their daily medications, especially if they have been evacuated. Because many people will have left home without their medicines, they will need help obtaining a supply of the medications that they take for chronic conditions. They may not know the names of their medications or the dosage. They may not be able to get in touch with their doctors, or their medical records may be destroyed. Some people may have to replace medications that require refrigeration. The rosters or computer backup may prove invaluable in these situations.

All pharmacies need a disaster plan that provides guidelines for acquiring and storing a supply of extra medication. This plan should be coordinated with the institution, community pharmacies, and state, regional, and federal agencies. The disaster plan should ensure that pharmaceuticals and supplies are part of the mutual aid agreement of an institution. Key components of a disaster plan for a pharmacy include:

A pharmacy technician prepares prescriptions at a retail pharmacy. During emergencies, pharmacy technicians assist in many functions that deal with preparing and distributing medicines, vaccines, and other agents.

- Providing services between the initial incident and the arrival of federal resources.
- Preparing lists of emergency contacts with contact information.
- Providing redundant communication within the pharmacy and within the institution.
- Preserving refrigerated materials if power is lost.
- Coordinating with vendors for ordering and delivery after the disaster.

- Delivering pharmaceuticals to alternate care sites.
- Providing medications to patients being transferred.
- Preparing procedures for receipt of medications/supplies from multiple sources.

If state laws permit, pharmacies can keep written instructions from physicians permitting them to issue prescribed medications if there is a disaster. Because some insurance plans limit the supply of medication dispensed to patients each month, pharmacies also need to work out in advance a plan with insurers about dispensing medications if patients have already received their monthly supply.

Hospitals and chain pharmacies can stockpile common medications. Because supplies in independent pharmacies may become quickly depleted, they should arrange with a distributor to obtain new supplies quickly in the event of a disaster. One of the responsibilities of a pharmacy technician is to track dates and rotate stockpiled supplies to ensure that medications have not expired.

Pharmacy technicians can perform an important role if a disaster calls for the large, rapid administration of medications, vaccines, or anti-infectives at a POD site.

If a pharmacy is in the path of a disaster, it may become a **mobile pharmacy**. This means that some pharmacists may choose to set up a small pharmacy in another place and bring with them a supply of the most commonly used medications. The pharmacy technician can tell callers where the mobile pharmacy is located, help the pharmacist organize and make the move, and set things up at the new location.

Other tasks pharmacy technicians can do include:

- Communicate with doctors about their supplies so that patients can continue to get their medications.
- Assess needs and ensure that there is a sufficient supply of medications.
- Call pharmacists in another state and transfer a prescription to that pharmacy.

A Closer Look

National Pharmacist Response Teams

Pharmacists, pharmacy students, and pharmacy technicians can help the federal government respond to large-scale disasters by joining one of ten **National Pharmacist Response Teams** that exist across the country. Volunteers receive training and equipment. While volunteers are working with the team, they are employees of the federal government. The purpose of the teams is to quickly distribute and administer anti-infectives, vaccines, or medications in case of a terrorist attack or other disaster. For information on volunteering for a team, contact the American Society of Health-System Pharmacists or go to the website at www.acpe-accredit.org/.

- Repackage stockpiled medications or reconstitute vaccines.
- Establish and maintain patient profiles.
- Take inventory and keep in stock prescription and over-the-counter medications.

Physical Therapist Assistants

Physical therapist assistants can help move patients safely and comfortably during an evacuation. They are familiar with the physical capabilities of patients and can advise on the types of assistance that patients will need during evacuations. During the response to Hurricane Katrina, physical therapist assistants gave massages to overworked volunteers, which helped them to continue working effectively during that stressful and physically challenging time. Physical therapy assistants can also be helpful in special needs shelters assisting evacuees with activities of daily living.

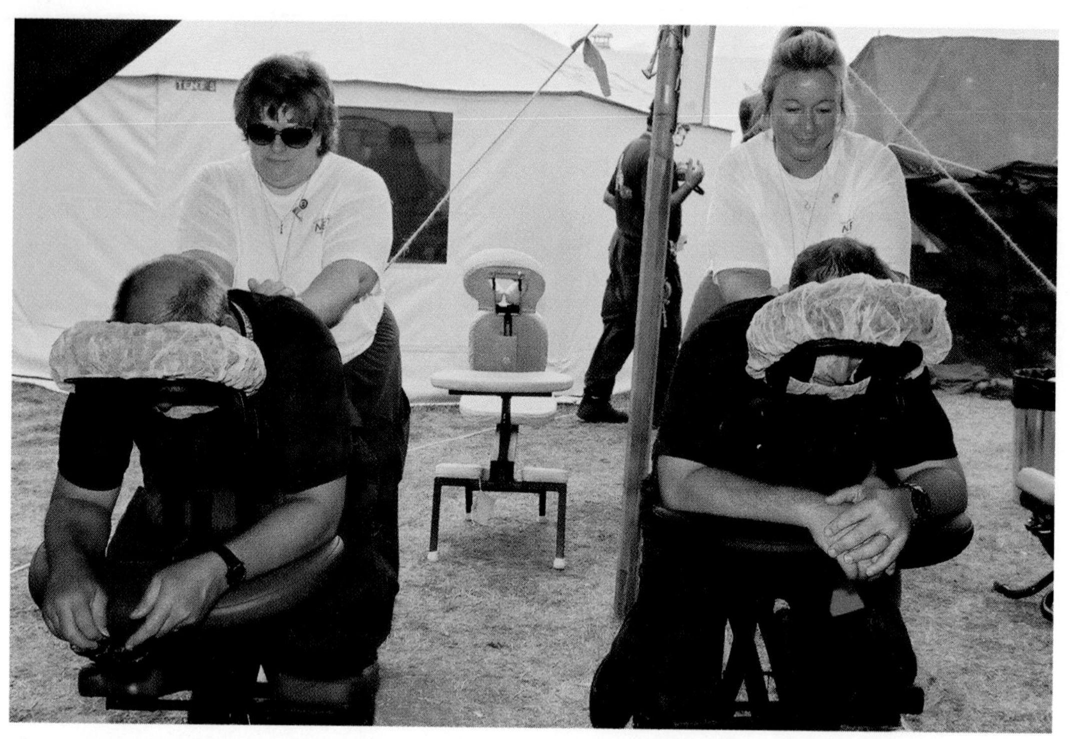

Physical therapist assistants can participate in disaster response, as shown in this photo, where rescue worker take a break from their work at the Pentagon, following the attack on September 11, 2001.

Respiratory Therapists

Respiratory therapists assist patients with breathing problems. Working under the direction of a physician, a respiratory therapist assesses patients, recommends treatments, and manages ventilators and artificial airway devices.

Because of their skills in airway management and ventilator support, respiratory therapists are critical in the initial efforts to resuscitate and stabilize patients. Additionally, they can aid in assessing and treating patients throughout the entire emergency medical response. Respiratory work in emergencies can involve operating and monitoring respiratory equipment; administering medical gases, aerosols, and respiratory medications; maintaining clearance of the natural and artificial airways of a patient; obtain-

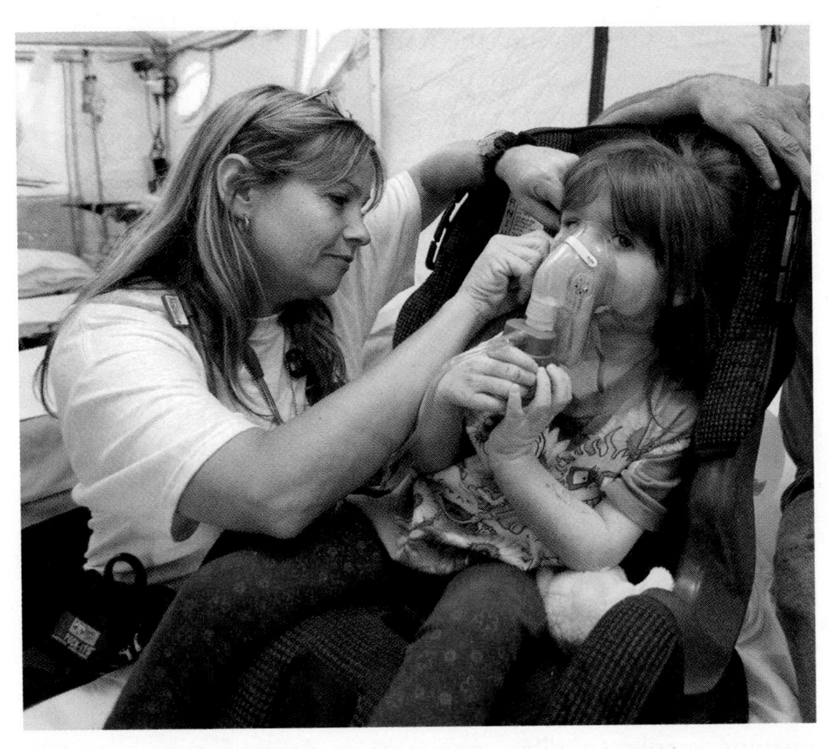

A respiratory therapist assists a young patient with her respirator mask. This scene was inside the Disaster Medical Assistance Team tent at Punta Gorda, Florida, following Hurricane Charley on August 2004.

ing blood samples and interpreting blood gas data; and providing primary assistance in CPR.

A respiratory therapist should be prepared to maintain ventilator support for patients, even if power fails. Therapists should know whom to contact for backup power and should notify that person if power fails. They should also be able to "bag," which means they manually provide ventilation to patients until the power returns. Additionally, respiratory therapists need to be trained on all types of mechanical ventilators because they may be moved to a facility that uses ventilators that are different from those at their regular facility.

In emergencies involving inhalation of toxic materials, a medical response would involve extensive resuscitation and respiratory support. There might be a rapid administration of antidotes. In a pandemic flu situation, the resources of hospitals may well be overwhelmed by a surge in patients.

Respiratory therapists can also:

- Resuscitate and stabilize patients requiring respiratory care.
- Move intubation and other airway management equipment, supplemental oxygen, and ventilators.
- Evacuate a person on a ventilator.
- Maintain airway and ventilation equipment.

Respiratory care providers may be called on to work in a variety of settings. For example, in special needs shelters they can handle respiratory problems and manage patients on oxygen. They may be asked to work on medical hotlines.

Surgical Technologists

Surgical technologists prepare operating rooms by selecting and opening sterile supplies, preparing patients for surgery, and passing instruments and sponges to surgeons. They can support emergency response in many support roles and in their usual roles in hospitals and emergency departments.

Surgical technologists are part of many DMAT teams. They receive additional training and can prepare casts and splints, help with surgical procedures in or out of the hospital, and provide wound care. Surgical technologists who are members of DMAT teams have a variety of work experience. They may come from hospital emergency departments, surgery departments, and orthopedics units. After the September 11, 2001 attacks, surgical technologists were part of the special burn teams sent to New York City to augment the city burn units.

Summary

The skills of all allied health professionals are invaluable in emergency response. Each profession can make an important contribution. Because emergency response is a team effort, it is helpful to know how different professions apply their knowledge and skills in an emergency.

Key Terms

mobile pharmacy

paramedic

National Laboratory Training Network (NLTN)

National Pharmacist Response Team

Reinforcing Terminology

Choose the word or phrase from the list of key terms that best matches each of the following descriptions.

1. Provides information for testing of suspicious or chemical agents ________
2. Provides anti-infectives in the event of a terrorist attack ________
3. Provides advanced life support on the way to a hospital ________
4. Dispensing medications in a temporary location ________

Reviewing Concepts

Choose the response that best answers the question or completes the sentence.

1. Which allied health professional conducts tests to help detect, diagnose, and treat diseases?
   a. emergency medical technician
   b. pharmacy technician
   c. laboratory technician
   d. health information technician

2. What is a role of physical therapist assistants during a disaster response?
   a. help evacuees with activities of daily living
   b. assist patients with breathing problems
   c. take patient histories
   d. distribute medicine and vaccines to evacuees

3. In a disaster response, which health professional helps to identify dead bodies?
   a. respiratory therapist
   b. radiation therapist
   c. physical therapist assistant
   d. surgical technologist

4. Which health professional provides wound care and prepares splints and casts?
   a. physical therapist assistant
   b. medical assistant
   c. surgical technologist
   d. medical imaging therapist

5. Which health professional has the most training to assist people at the scene of a disaster?
   a. laboratory technician
   b. surgical technologist
   c. medical assistant
   d. emergency medical technician

6. What skills are helpful to a medical assistant?
   a. computer repair skills
   b. mechanical ventilator skills
   c. advanced life support skills
   d. administrative and clerical skills

7. An emergency medical technician can administer
   a. POD.
   b. CT.
   c. CPR.
   d. ALS.

8. What role can a health information technician play during an emergency?
   a. help victims obtain a supply of medications
   b. keep patient records from being destroyed
   c. schedule and receive patients at a shelter
   d. take X-rays to determine injuries

9. What should be part of the emergency preparedness activities of a pharmacy technician?
   a. making backup copies of customers' medication records
   b. conducting routine maintenance of the electric-powered medical devices of customers
   c. supplying insurance information for go bags of customers
   d. helping customers preregister for a special needs shelter

10. Which skilled professional is critical in initial efforts to resuscitate victims?
    a. surgical technologist
    b. radiation therapist
    c. respiratory therapist
    d. physical therapist

Exploring Emergency Preparedness Issues

1. Contact your state or national professional association. Find out about the activities that they have planned for your profession in the event of a disaster and discuss this information in several paragraphs.
2. Make an "I Can Do" list. Include the tasks that you can perform if you are working at your regular job, volunteering as part of a team, or volunteering in the community.
3. Which of your healthcare colleagues are you likely to work with during a disaster? What similar tasks will each of you work on together? Which different tasks will each of you work on? Summarize your response in one page.

Websites

Emergency preparedness is a complex and evolving field. As more and more professionals in the healthcare community take an active role, there will continue to be new things to learn. As you learn more about your role in emergency preparedness, you will need to stay current. You can find helpful information about emergency preparedness at the websites listed below.

Federal Resources

http://www.bt.cdc.gov/

The Centers for Disease Control and Prevention (CDC) website provides essential information about the health effects of disasters and the important activities that healthcare professionals can undertake to ensure that their organization is well prepared and can respond effectively. The Preparation and Planning section under "More Topics and Resources" contains links to learning more about the readiness activities of many groups.

http://www.bt.cdc.gov/lrn/factsheet.asp

This fact sheet by the Centers for Disease Control and Prevention (CDC) provides a description about the Laboratory Response Network and how it works.

http://www.bt.cdc.gov/stockpile/

This Centers for Disease Control and Prevention (CDC) website provides details about the Strategic National Stockpile and how to obtain medicines and supplies from the stockpile.

http://www.dhs.gov/index.shtm

The Department of Homeland Security (DHS) website provides information about the federal response and recovery efforts. Explore this website to better understand how the DHS is organized and what it is prepared to do in a disaster.

http://www.dhs.gov/xprepresp/training

This section of the website of the Department of Homeland Security provides information on training, technical assistance, and exercises. It also provides links to key federal sources for additional preparation.

http://www.epa.gov

In their role of protecting human health and the environment, the Environmental Protection Agency (EPA) is often called in when a disaster results in environmental damage or the releases of hazardous materials. Their website provides important information about items as diverse as cleaning up garbage to the disinfection of drinking water during emergencies.

http://www.fema.gov/

The Federal Emergency Management Agency (FEMA) website is a very comprehensive site about disasters. It provides information about many types of disasters. Go to this website to learn about actions you can take before and after a disaster to protect yourself and your family. This website also explains the process of getting assistance after a disaster.

http://www.hhs.gov/disasters/discussion/responders/

The Department of Health and Human Services provides key information and resources for healthcare professionals to help them prepare and respond to man-made and natural disasters. Of particular value are their web links to other federal websites that contain detailed information on lab safety and personal protection.

http://mentalhealth.samhsa.gov/cmhs/EmergencyServices/

The Center for Mental Heath Services works to ensure that disaster victims receive immediate, short-term, and/or crisis counseling, and ongoing support. This website provides specific information about the mental health impacts of disaster and what mental health workers do to help disaster victims.

http://www.nws.noaa.gov/

To learn more about the weather in your area, go to the website of the National Weather Service. At this website, you can track storms that may be forming and know whether your community is at risk.

http://www.ready.gov/

The Ready.Gov website provides information about contacting your local government to learn more about how you can help your community prepare. This website also provides useful information about putting together emergency kits.

Non-governmental Organizations

http://www.cbsnews.com/digitaldan/disaster/disasters.shtml

The CBS News Disaster Links is the most comprehensive list of web links about disaster topics on the Web. In one click, you can get connected to the major sources of information about every kind of disaster. Regular exploration of this website will help you stay informed about news events and preparedness activities across the United States.

http://www.jointcommission.org/

The Joint Commission has actively required that healthcare organizations prepare for emergencies. Their website provides information about their accreditation requirements.

As healthcare professionals, you should become familiar with the Joint Commission. Explore this website. To learn more about their work in preparedness, type "emergency preparedness" in the search box.

http://www.redcross.org/

The American Red Cross is a key nongovernmental agency in disaster response. Their website provides heartwarming stories about the real life experiences of disaster victims. It is also a good resource for important background information about preparing for disasters.

Worker Safety

To ensure that you can protect yourself during a disaster and in the days and weeks that follow during recovery, learn more about the protections that your employer is required to provide.

http://www.cdc.gov/niosh/topics/emres/

The National Institute for Occupational Safety and Health (NIOSH) website provides links to important safety and health information to guide first responders on managing a disaster site and the use of personal protective equipment.

http://www.osha.gov/dts/osta/bestpractices/firstreceivers_hospital.html

The Occupational Safety and Health Administration (OSHA) website provides practical information to assist hospitals in protecting the personnel who work in the emergency department during a disaster involving contaminated patients. This website provides information about decontamination, personal protective equipment, and employee training.

Volunteering

http://www.hrsa.gov/esarvhp/

To learn more about volunteering, go to the website of the Emergency System for Advance Registration of Volunteer Health Professions (ESAR-VHP).

http://www.nvoad.org/

To learn more about nonmedical opportunities to volunteer in a disaster, go to the website of the National Voluntary Organizations Active in Disaster (NVOAD).

Professional Associations

http://www.ama-assn.org/ama/pub/category/6206.html

Through the Center for Public Health Preparedness and Disaster Response, the American Medical Association (AMA) provides materials for doctors and other health professionals. This website provides important educational resources for healthcare professionals across the United States.

http://www.ama-assn.org/ama/pub/category/12606.html

The National Disaster Life Support Program of the American Medical Association (AMA) provides courses to help healthcare professionals understand their roles in the broader disaster response system. The courses include Core Disaster Life Support, an introduction to all-hazards preparedness; Basic Disaster Life Support, a review of the all-hazards approach, including information on the role of the healthcare professional; and Advanced Disaster Life Support, an advanced course for those who have taken the basic course.

http://www.aphl.org/programs/emergency_preparedness/Pages/default.aspx

The Association of Public Health Laboratories (APHL) has actively organized resources for all lab professionals. Their website includes critical information about Laboratory Response Network activities and handling all hazards, including bioterrorism agents.

http://www.aarc.org/community/disaster_roundtable.asp

The American Association for Respiratory Care (AARC), the professional society for respiratory therapists, has formed a disaster response roundtable website. Through the roundtable website, you can be linked to members interested in sharing information about preparedness and response in times of disaster. At this website, you can join an electronic mailing list and network with other people who are interested in preparedness. You can also locate the lead person in your state.

http://www.ashp.org/s_ashp/cat1c.asp?CID=505&DID=547

The American Society of Health System Pharmacists maintains a Resource Center to provide pharmacists with information and guidance on the emergency preparedness and response activities across the United States. To learn about the specific tasks which pharmacists are required to prepare for, click on "Leader's Checklist" under Pharmacists' Roles and Responsibilities.

http://www.getreadyforflu.org/preparedness/influenza_main.htm

The American Public Health Association provides important information through their *Get Ready Campaign* about getting prepared for pandemic influenza or any other emerging infectious disease. Their materials are clearly and simply written so that individuals, families, and communities can easily know what to do to get prepared and where to go for more information.

http://www.naemt.org/

The National Association of Emergency Medical Technicians is the professional group dedicated to emergency management services knowledge, education, and professional development.

http://www.caahep.org

CAAHEP, the Commission on Accreditation of Allied Health Education Programs, is working with 19 health professions on emergency-preparedness standards for educational programs. CAAHEP has compiled information and web resources for educators; click on Emergency Preparedness Project.

Vulnerable Populations And Disabilities

http://www.ada.gov/emergencyprep.htm

The Americans for Disabilities Act (ADA) of 1990 requires that governments, organizations, and businesses in the United States ensure that people with disabilities receive the same protection in emergencies as the rest of the population. This webpage has links to the detailed guide, *An ADA Guide for Local Governments: Making Community Emergency Preparedness and Response Programs Accessible to People with Disabilities*. This guide has drawings and describes how to make accommodations in preparation for and during an emergency.

http://www.bt.cdc.gov/workbook/

The Centers for Disease Control and Protection (CDC) has developed *The Public Health Workbook to Define, Locate and Reach Special, Vulnerable and At-Risk Populations in an Emergency* to help agencies that serve vulnerable populations in locating these populations in an emergency. The workbook includes checklists, a sample telephone survey, information on working with others, and more.

http://www.cdc.gov/diabetes/news/docs/hurricanes.htm

In response to Hurricane Katrina in 2005, the Centers for Disease Control and Prevention (CDC) developed a website called "Help for People with Diabetes Affected by Hurricanes." This valuable resource is appropriate to other disaster settings. It provides information about insulin, drugs, and equipment; specific health advice; drug resources for evacuees; and health coverage.

http://www.cepintdi.org/

The Community Emergency Preparedness Information Network website is a project of the Telecommunications for the Deaf and Hard of Hearing (TDI). This website educates emergency personnel about issues that impact people with hearing loss during an emergency or disaster and provides education and information about emergency preparedness to people who are deaf or hard of hearing.

http://www.diabetes.org/for-parents-and-kids/living-with-diabetes/emergency-preparedness.jsp

The American Diabetes Association provides information to help diabetics prepare for emergencies.

http://www.jik.com/disaster.html

June Isaacson Kailes is a leading disabilities expert. Her website is an excellent resource on disability and emergency preparedness.

http://www.medicare.gov/Publications/Pubs/pdf/10150.pdf

This webpage contains *Preparing for Emergencies: A Guide for People on Dialysis,* which describes the steps to get ready for an emergency. Available in English and Spanish, this guide provides direction on gathering medical information, making alternative arrangements for treatment, preparing an emergency stock of medications and food, disinfecting after, and what to do if on a dialysis machine in an emergency.

http://www.preparenow.org/

PrepareNow.org is an organization in California that supports the needs of vulnerable populations and people with disabilities during a disaster. Go to *http://www.preparenow.org/tipcrd.html* to find tips on creating an emergency health information card. Go to *http://www.preparenow.org/doctip.html* to find tips for collecting emergency documents.

http://www.remm.nlm.gov/remm_FirstResponder.htm

This website provides important information on radiological management for first responders in the field. It includes triage, medical management, the use of personal protective equipment, casualty management, information about exposure to radiation, and study courses.

Self-study Resources

Use these websites to learn more by taking courses in which you study at your own pace.

http://training.fema.gov/IS/crslist.asp

The Emergency Management Institute (EMI) is part of the Federal Emergency Management Agency. EMI has courses onsite in Emmitsburg, Maryland, and also provides an independent study program that is available online. EMI offers a broad range of courses about emergency response and recovery. Academic credit can be earned for many of the courses through Frederick Community College in Frederick, Maryland.

http://training.fema.gov/emiweb/is/is700.asp

To learn more about the National Incident Management System (NIMS), take the three-hour introductory course. The course covers the principles and key components of NIMS. The course also provides an opportunity to do some planning, which you can print and keep.

Abbreviations

ALS	advanced life support
CBRN	chemical, biological, radiological, and nuclear
CDC	Centers for Disease Control and Prevention
CERT	Community Emergency Response Team
CISD	critical incident stress debriefing
CPR	cardiopulmonary resuscitation
CT	computer tomography
DMAT	Disaster Medical Assistance Team
DMORT	Disaster Mortuary Operations Response Team
EMAC	Emergency Management Assistance Compact
EMS	emergency medical service
EMT	emergency medical technician
EOC	emergency operations center
EOP	emergency operations plan
ESAR-VHP	Emergency Systems for Advance Registration of Volunteer Health Professionals
FBI	Federal Bureau of Investigation
FEMA	Federal Emergency Management Agency
HICS	Hospital Incident Command System
ICS	Incident Command System
IV	intravenous
TJC	The Joint Commission
LEMA	Local Emergency Management Agency
LRN	Laboratory Response Network
MOU	memorandum of understanding
MR	magnetic resonance

MRC	Medical Reserve Corps
NDMS	National Disaster Medical System
NIMS	National Incident Management System
NRF	National Response Framework
OSHA	Occupational Safety and Health Administration
POD	Point of Distribution
PPE	personal protective equipment
SARS	severe acute respiratory syndrome
SNS	Strategic National Stockpile
START	Simple Triage and Rapid Treatment

References and Resources

Chapter 1

Becker, Cindy and Joseph Mantone. "Cries for Help: Hospitals Overwhelmed By Katrina's Aftermath." *Modern Healthcare,* September 5, 2005, pp. 8-10.

Blake, P. A. "Communicable Disease Control." In: Gregg, M. B., ed. *The Public Health Consequences of Disasters.* (Atlanta, GA: U.S. Centers for Disease Control and Prevention, Department of Health and Human Services, 1989), pp. 7-12.

Centers for Disease Control and Prevention. "Assessment of Health-Related Needs After Hurricanes Katrina and Rita." *MMWR: Morbidity and Mortality Weekly Report,* Vol. 55, January 20, 2006, pp. 38-41.

Centers for Disease Control and Prevention. "Public Health Response to Hurricanes Katrina and Rita." *MMWR: Morbidity and Mortality Weekly Report,* Vol. 55, January 20, 2006, pp. 29-30.

Landesman, L. Y. *Public Health Management of Disaster: The Practice Guide.* 2nd Edition (Washington, D.C.: American Public Health Association, 2005), pp. 1-31.

Lister, S. A. "The Public Health and Medical Response." *CRS Report for Congress,* September 21, 2005.

Rudowitz, Robin, Diane Rowland, and Adele Shartzer. "Health Care in New Orleans Before and After Hurricane Katrina." *Health Affairs,* August 29, 2006. Retrieved 5/19/07 (*http://content.healthaffairs.org/cgi/content/full/hlthaff.25.w393v1/DC1*).

Scheld, Michael, William A. Craig, and James M. Hughes. *Emerging Infections 4.* (Washington, D.C.: ASM Press, 2000), pp. 137-147.

House of Representatives, Select Bipartisan Committee to Investigate the Preparation for and Response to Hurricane Katrina. *A Failure of Initiative: Final Report of the Select Bipartisan Committee to Investigate the Preparation for and Response to Hurricane Katrina.* 109th Congress, Second Session. (Washington, D.C.: U.S. Government Printing Office, 2006), pp. 267-308.

Sternberg, Steve. "New Orleans Deaths Up 47%." *USA Today,* Fri/Sat/Sun, June 22-24, 2007, p. 1a.

Chapter 2

Barbera J. A. and Macintyre A. G. Medical and Health Incident Management (MaHIM) System Final Report, 2002. Retrieved February 22, 2007 (*www.gwu.edu/~icdrm/publications/MaHIMV2finalreportsec2.pdf*).

The CNA Corporation. *Medical Surge Capacity and Capability: A Management System for Integrating Medical and Health Resources During Large-Scale Emergencies.* (Alexandria, VA: The CNA Corporation, August 2004). Retrieved February 22, 2007 (*www.cna.org/documents/mscc_aug2004.pdf*).

Federal Emergency Management Agency, U.S. Department of Homeland Security. *National Incident Management System Chapter II: Command and Management.* March 1, 2004, pp. 8-10. Retrieved 2/17/07 (*www.nimsonline.com/nims_3_04/preparedness.htm*).

Giordano, Lorraine. "Terror in the Towers: A Case Study of the World Trade Center Bombing," in Landesman, Linda Young (ed.) *Emergency Preparedness in Healthcare Organizations*, 1996, The Joint Commission of Healthcare Organizations, Oakbrook Terrace, IL. pp. 109-121.

The Institute for Crisis, Disaster and Risk Management, George Washington University. *Emergency Management Principles and Practices for Healthcare Systems.* (Washington, D.C.: U.S. Department of Veterans Affairs, June 2006), pp. 1-1–1-72, 2-70–2-95.

Joint Commission on Accreditation of Healthcare Organizations. *2006 Hospital Accreditation Standards for Emergency Management Planning and Emergency Management Drills.* JCAHO Standards EC.4.10 and EC.4.20," 2007.

Joint Commission on Accreditation of Healthcare Organizations. *2006 Long Term Care Accreditation Standards for Emergency Management Planning and Emergency Management Drills.* JCAHO Standards EC.4.10 and EC.4.20, 2007.

Landesman, L. Y. *Public Health Management of Disaster: The Practice Guide.* 2nd Edition (Washington, D.C.: American Public Health Association, 2005), pp. 33-89.

Lindell, M. K., C. Prater, and R. W. Perry. *Introduction to Emergency Management.* (Hoboken, NJ: John Wiley & Sons, Inc., 2007), pp. 11-16, 23-30, 281-292.

Chapter 3

American Red Cross. *How Do I Become a Volunteer in Disaster and Client Services?* Retrieved February 22, 2007 (*www.oc-redcross.org/show.aspx?mi=2940*).

The Citizens Corps "Community Emergency Response Team (CERT)" Retrieved 10/14/2007 (*www.citizencorps.gov/cert/*).

Ehrenreich, John H. and Sharon McQuade. "Coping with Disaster: A Guide for Relief Workers." In: *Coping with Disasters: A Guidebook to Psychological Intervention.* (NY: State University of New York, 2001). (*www.mhwwb.org/CopingWithDisaster.pdf*).

Federal Emergency Management Agency and American Red Cross. *Family Disaster Plan.* (*www.redcross.org/services/disaster/0,1082,0_601_,00.html*).

Health Resources and Services Administration, U.S. Department of Health and Human Services. *Emergency System for Advance Registration of Volunteer Health Professionals.* (*www.hrsa.gov/esarvhp/default.htm*).

Hodge, James G. *Legal and Regulatory Issues Concerning Volunteer Health Professionals in Emergencies: An Overview.* The Center for Law and Public Health at Georgetown and John Hopkins Universities. Presentation at the Partners Breakfast, Annual Meeting of the American Public Health Association, Boston Massachusetts, November 7, 2006.

Medical Reserve Corps, "About Volunteering." (*www.medicalreservecorps.gov/AboutVolunteering*).

Young, Bruce H., et al. *Disaster and Trauma: A Guidebook for Clinicians and Administrators.* (White River Junction, VT: The National Center for Post Traumatic Stress Disorder, VA Medical and Regional Office Center, 2001).

Chapter 4

Agency for Healthcare Research and Quality. *Development of Models for Emergency Preparedness: Personal Protective Equipment, Decontamination, Isolation/Quarantine, and Laboratory Capacity.* AHRQ Publication No. 05-0099, (Rockville, MD: Agency for Healthcare Research and Quality, August 2005). Retrieved March 24, 2007 (*www.ahrq.gov/research/devmodels/*).

Agency for Toxic Substances and Disease Registry. *Primer on Health Risk Communication Principles and Practices, June 2001.* (*www.atsdr.cdc.gov/HEC/primer.html.*)

Association of Public Health Laboratories. "Ready, Set Respond: Chemical Terrorism Preparedness in the Nation's State Public Health Laboratories." *Public Health Laboratory Issues in Brief,* March 2006. (*www.aphl.org/Documents/Global_docs/chemical_terrorism_3-06.pdf*).

Barbera, Joseph, et al. "Large Scale Quarantine Following Biological Terrorism in the United States." *JAMA: Journal of the American Medical Association,* Vol. 286, 2001, pp. 2711-2717.

California Association of Environmental Health Specialists. *Disaster Field Manual for Environmental Health Specialists.* (Carmichael, CA: California Association of Environmental Health Administrators, 1998), pp. 9-1–9-10.

Center for Mental Health Services. *Communicating in a Crisis: Risk Communication Guidelines for Public Officials.* (Washington, D.C.: Substance Abuse and Mental Health Services, U. S. Department of Health and Human Services, 2002).

Centers for Disease Control and Prevention. "Cardiac Deaths After A Mass Smallpox Vaccination Campaign—New York City, 1947." *MMWR: Morbidity and Mortality Weekly Report,* Vol. 52, No. 39, October 3, 2003, pp. 933-936.

Centers for Disease Control and Prevention. *Chemical Agents: Facts About Personal Cleaning and Disposal of Contaminated Clothing.* (Atlanta, GA: Centers for Disease Control and Protection, U.S. Department of Health and Human Services, August 16, 2006). (*www.bt.cdc.gov/planning/personalcleaningfacts.asp*).

Centers for Disease Control and Prevention. *Clean Hands Save Lives: Emergency Situations.* (Atlanta: GA: Centers for Disease Control and Prevention, U.S. Department of Health and Human Services, 2007) (*www.bt.cdc.gov/diasters/pdf/handhygienefacts.pdf*).

Garner, Julia. "Guidelines for Infection Control Practices in Hospitals." *Infection Control and Hospital Epidemiology,* Vol. 17, 1996, pp. 53-80.

Landesman L.Y. *Public Health Management of Disasters*: The Pocket Guide. (Washington D.C.: American Public Health Association, 2006) pp. 78- 81.

National Association of County and City Health Officials. *Issues to Consider: Isolation and Quarantine.* January 2006. (*http://biotech.law.lsu.edu/cases/pp/naccho-quarantine.pdf*).

National Fire Protection Association. "Emergency Preparedness: NFPA Standards," Quincy, MA: National Fire Protection Association). (*www.nfpa.org/catalog/category.asp?category%5Fname=Emergency+Preparedness+%7C+Codes%2FStandards&Page=1&src=catalog*).

National Institute for Occupational Safety and Health. *Fact Sheet for Workers in Secondary Response and Other Supporting Roles at the World Trade Center.* (Atlanta, GA: Centers for Disease Control and Protection, U.S. Department of Health and Human Services, September 2001). (*www.cdc.gov/niosh/erfaqs.html*).

National Institute for Occupational Safety and Health. *NIOSH Pocket Guide to Chemical Hazards (NPG),* DHHS (NIOSH) Publication No. 2005-149. (Atlanta, GA: Centers for Disease Control and Prevention, U.S. Department of Health and Human Services, September 2005).

New York City Department of Health. *Guidance Document for Development of Protocols for Management of Patients Presenting to Hospital Emergency Departments and Clinics with Communicable Diseases of Urgent Public Health Concern.* 2005. (*www.nyc.gov/html/doh/downloads/word/bhpp/bhpp-train-emergency-guidance-01.doc*).

Occupational Safety and Health Administration. "Bloodborne Pathogens." *Code of Federal Regulations.* Standard Number 1910.1030. (*www.osha.gov/pls/oshaweb/owadisp.show_document?p_table=STANDARDS&p_id=10051*).

Occupational Safety and Health Administration. *How to Plan for Workplace Emergencies and Evacuations.* (Washington, D.C.: Occupational Safety and Health Administration, U.S. Department of Labor, 2001). (*www.osha.gov/Publications/osha3088.pdf*).

Oklahoma Department of Civil Emergency Management. *After Action Report, Alfred P. Murrah Federal Building Bombing, 19 April 1995, Oklahoma City, Oklahoma.* Retrieved May 5, 2007 (*www.ok.gov/OEM/documents/BombingAfterActionReport.pdf*).

Pierce, Ronald. "Hospital Pandemic Preparedness: Anticipating Unprecedented Surge of Infectious Waste." *Medical Waste Management,* Vol. 2, No. 3, July-Sept 2006, pp. 1-4.

Powers, Michael J. "Incident Public Information and Crisis Communications." *What Should We Know? Whom Do We Tell? Leveraging Communication and Information to Counter Terrorism and Its Consequence.* (Memorial Institute for the Prevention of Terrorism, December 2001). (*www.terrorisminfo.mipt.org/pdf/CBACI/LevComm/externalII.pdf).*

SARS Commission Final Report. *Disaster at Sunnybrook.* Volume Two, Spring of Fear, December 2006, pp. 393-412. Retrieved April 21, 2007 (*www.sarscommission.ca/report/v2-pdf/Vol2Chp3x.pdf*).

Smith, Gary J. *Floods: An Environmental Health Practitioners's Operational Field Guide, National Environmental Health Monographs, Counter Disaster Series No. 1, 1999.* (Queensland, South Australia: National Environmental Health Forum, Public and Environmental Health Service, Department of Human Services). Retrieved May 24, 2007 (*http://enhealth.nphp.gov.au/council/pubs/pdf/floods.pdf*).

Stopford, B. M., L. Jevitt, M. Ledgerwood, C. Singleton, M. Stolmack. "Chapter 3: Decontamination," *Development of Models for Emergency Preparedness: Personal Protective Equipment, Decontamination, Isolation/Quarantine, and Laboratory Capacity.* (Rockville, MD: Agency for Healthcare Research and Quality, U.S. Department of Health and Human Services, August 2005), pp. 69-97. (*www.ahrq.gov/research/devmodels/devmodels.pdf*).

Stopford, B.M., L. Jevitt, M. Ledgerwood, C. Singleton, M. Stolmack. "Chapter 4: Isolation/ Quarantine," In: *Development of Models for Emergency Preparedness: Personal Protective Equipment, Decontamination, Isolation/Quarantine, and Laboratory Capacity.* (Rockville, MD: Agency for Healthcare Research and Quality, U.S. Department of Health and Human Services, August 2005), pp. 101-115. (*www.ahrq.gov/research/devmodels/devmodels.pdf*).

U.S. Department of Labor. "HAZWOPER Standard, Paragraph 1910.120(q)(4))," *OSHA Best Practices for Hospital-Based First Receivers of Victims from Mass Casualty Incidents Involving the Release of Hazardous Substances.* January 2005, p. 17. (*www.osha.gov/dts/osta/bestpractices/firstreceivers_hospital.html*).

World Health Organization. *Chemical Terrorism in Japan: The Matsumoto and Tokyo Incidents.* (*www.opcw.org/resp/html/japan.html*).

Chapter 5

The Centers for Medicare and Medicaid Services. *Preparing for Emergencies: A Guide for People on Dialysis.* (Baltimore, MD: The Centers for Medicare and Medicaid Services, U.S. Department of Health and Human Services, November 2002). (*www.medicare.gov/Publications/Pubs/pdf/10150.pdf*).

Lam, Clarence, Richard Waldhorn, Eric Toner, Thomas Inglesby, and Tara O'Toole. "The Prospect of Using Alternative Medical Care Facilities in an Influenza Pandemic." *Biosecurity and Bioterrorism: Biodefense Strategy, Practice, and Science,* Vol. 4, No. 4, December 2006, pp. 384-390.

Lessons Learned Information Sharing. *Disabilities and Other Special Needs: The Aging Services Council of Central Texas's Emergency Supply Kits for Homebound Elderly Residents,* March 2007. (*https://www.llis.dhs.gov/member/secure/detail.cfm?content_id=23467*).

National Center for Chronic Disease Prevention and Health Promotion. *Chronic Diseases and Vulnerable Populations in Times of Natural Disaster: An Action Guide.* (Atlanta, GA: Centers for Disease Control and Prevention, Forthcoming).

Powell, Robyn and Sheldon. *The Impact of Hurricanes Katrina and Rita on People with Disabilities: A Look Back and Remaining Challenges.* (Washington D.C.: National Council on Disability, August, 2006).

Chapter 6

Benjaminov, Ofer, Miriam Sklar-Levy, Avraham Rivkind, Maya Cohen, Gabi Bar-Tal, and Michael Stein. "Role of Radiology in Evaluation of Terror Attack Victims." *AJR: American Journal of Roentgenology*, Vol. 187, September 2006, pp. 609-616.

Chin, Thomas W. F., Clarence Chant, Rosemary Tanzini, and Janice Wells. "Severe Acute Respiratory Syndrome (SARS): The Pharmacist's Role." *Pharmacotherapy,* Vol. 24, No. 6, 2004, pp. 705-712.

Cohen, Victor. "Organization of a Health-System Pharmacy Team to Respond to Episodes of Terrorism." *American Journal of Health System-Pharmacy,* Vol. 60, No. 12, 2003, pp. 1257-1263. (*www.medscape.com/viewarticle/456790*).

Ford, Stephen and John Grabenstein. "Pandemics Avian Influenza (H5N1) and a Strategy for Pharmacists." *Pharmacotherapy*, Vol. 26, No. 3, 2006, pp. 312-322.

Heffernan, Thomas E., Srinesh Alle, and Charles C. Matthews. "Weathering the Storm: Maintaining an Operational Radiology Department at Ochsner Medical Center Throughout Hurricane Katrina." *Radiology*, Vol. 242, 2007, pp. 334-337. Retrieved April 14, 2007 (*http://radiology.rsnajnls.org/cgi/content/full/242/2/334*).

Kyes, Kris. "Radiation Disaster Preparedness: Radiology's Role." *Imaging Economics,* September 2003. Retrieved April 14, 2007 (*www.imagingeconomics.com/issues/articles/2003-09_05.asp?mode=print&*).

Rando, Frank. *When Seconds Count: Preparing Respiratory Therapists for Mass Casualty Incident Response.* American Association of Respiratory Care. Retrieved April 14, 2007 (*www.aarc.org/headlines/rtsandmasscasualties.asp*).

Reese, Meghan. "Forensic Radiology: Dead Men Do Tell Tales." *Radiology Today,* Vol. 7, No. 18, September 11, 2006, p. 12. Retrieved March 29, 2007 (*www.radiologytoday.net/archive/rt09112006p12.shtml*).

Tang, Patrick, et al. "Interpretation of Diagnostic Laboratory Tests for Severe Acute Respiratory Syndrome: The Toronto Experience." *Canadian Medical Association Journal,* Vol. 170-1, January 6, 2004, pp. 47-54.

Teeter, David. "Disaster Preparedness and Pharmacy: An Important Partnership." *U.S. Pharmacist,* Vol. 29, No. 2, February 15, 2004. Retrieved April 14, 2007 (*www.uspharmacist.com/index.asp?show=article&page=8_1217.htm*).

Glossary

all-hazards approach An approach that provides general preparation and training that can be applied in a wide range of emergency situations.

autoclaving A method of medical waste treatment that exposes the waste to steam at sufficient temperature under a sufficient amount of pressure and for sufficient time to ensure that microorganisms are destroyed. Also called *steam sterilization*. See also *medical waste*.

biohazardous waste The medical waste consisting of body fluids or blood products and laboratory specimens from various procedures or biopsies, See also *medical waste*.

bioterrorism The deliberate release of viruses, bacteria, or other agents that can cause illness or death in people, animals, or plants. See also *terrorism*.

burnout A condition in which people suffer extreme emotional exhaustion. Symptoms include lack of enthusiasm, fatigue, irritability, difficulty concentrating or sleeping, and a variety of physical symptoms such as headaches or gastrointestinal upsets.

chain of command A system in which reporting relationships are clarified thereby eliminating the confusion caused by multiple, conflicting directives.

chronic diseases The diseases that are long lasting and treatable, but not curable such as heart disease and diabetes.

cohorted patients Patients with similar symptoms.

community A jurisdiction such as a town, city, metropolitan area, or county.

Community Emergency Response Team A program for volunteers that educates people about disaster preparedness and trains them in basic disaster response skills, such as fire safety, light search and rescue, and disaster medical operations.

comprehensive emergency management A system that provides for carrying out the functions of emergency management and encompasses four phases: mitigation, preparedness, response, and recovery. See also *mitigation, preparedness, response*, and *recovery*.

contagious A disease that can be transmitted from person to person.

contamination The accidental release of hazardous substances that places people at risk.

continuity of care The ongoing, coordinated care from the same group of healthcare professionals when needed. See also *chronic diseases*.

credentialed The verification of identity, license, training, skills, and competencies for volunteer workers.

crisis intervention A set of techniques that helps disaster workers feel more in control during a crisis. Crisis intervention involves individual or group counseling.

critical incident stress debriefing A technique used to help a group of disaster workers process their disaster experience.

decontamination The removal or neutralization of harmful substances such as chemicals, agents or radioactive materials.

defusing A brief, informal procedure used to help disaster workers deal with their reactions to events. Mental health professionals hold 10- to 30-minute sessions in which workers can express their thoughts and feelings about what they are doing.

disaster An emergency that requires outside assistance. See also *man-made disaster* and *natural disaster*.

Disaster Medical Assistance Team (DMAT) A group of medical professionals and support staff sent by the National Disaster Medical System to supplement local medical resources.

Disaster Mortuary Operations Response Team (DMORT) A group sent by the National Disaster Medical System to provide victim identification and mortuary services. Teams include funeral directors, medical examiners, pathologists, medical records technicians, dental assistants, radiographers, mental health specialists, and support staff.

droplet precautions The protective procedures followed by health professionals when they are examining a patient with symptoms of a respiratory infection, particularly if fever is present. Includes wearing a medical mask.

emergency An incident that threatens public safety, health, and welfare. Some emergencies are limited, and others require a more rigorous response. In a hospital, the term *emergency* is usually defined as a situation in which an individual patient requires immediate care or a high level of care. The field of emergency preparedness defines emergency as a broader incident involving groups of people.

emergency drills The simulated emergencies designed to identify the strengths and weaknesses of an emergency plan of a hospital.

Emergency Management Assistance Compact (EMAC) A mutual aid agreement that facilitates the sharing of resources, personnel, and equipment across state lines during times of disaster. Applies to the 50 states, the District of Columbia, Puerto Rico, and the U.S. Virgin Islands.

emergency medical technician (EMT) A health professional who is trained in ambulance operations and procedures.

emergency operations center (EOC) A place where leaders of an emergency operation gather to coordinate a response.

emergency operations plan The written procedures and protocols for responding to emergencies.

emergency preparedness The act of making plans to prevent, respond to, and recover from emergencies.

emergency response A coordinated action to meet the needs of communities affected by an emergency.

emergency response providers The people such as public safety workers, police officers, fire fighters, and emergency medical technicians and paramedics who are trained in advance to respond to emergencies.

Emergency System for Advance Registration of Volunteer Health Professionals (ESAR-VHP) The system designed to assess and preregister volunteer health professionals to work within their own states and in other states during public health and other emergencies. Each state-based system contains readily available, verifiable, current information about the identity, licensing, and credentials of a volunteer.

emerging infectious disease A new or previously limited infection that spreads rapidly in the population.

evacuation The organized, phased, and supervised withdrawal of people from a dangerous or potentially dangerous area.

Federal Emergency Management Agency (FEMA) A part of the U.S. Department of Homeland Security and the lead federal agency during any national emergency. FEMA has the lead responsibility for coordinating a regional response and providing management and response staff. FEMA is also in charge of providing shelter, food, and first aid to victims.

first aid The simple, potentially life saving medical techniques delivered with minimal or no medical equipment.

fit-tested A procedure to ensure that the make, model, style, and size of respirator are appropriate for the wearer. See also *respirator*.

flash flooding A flooding that occurs suddenly.

go bag A small collection of essentials that individuals and families may need in the event of an evacuation.

hazards The risks or dangers that have the potential to harm people, property, and the natural environment.

hospital emergency operations plan A plan designed for hospitals to deal with emergencies. See also *emergency operations plan.*

Hospital Incident Command System (HIPS) A system that clarifies roles and responsibilities and enables facilities to handle an emergency without delay by establishing a chain of command and specifying staff responsibilities. See also *Incident Command System.*

Incident Command System A management system that organizes and coordinates emergency response.

incident commander The leader of the Incident Command System. Assigns responsibilities and communicates the status of the response to the emergency operations center. See also *Incident Command System* and *emergency operations center.*

incineration A method of disposal by burning medical waste at temperatures ranging from 1,800°F to 2,000°F (982°C to 1,093°C). See also *medical waste.*

influenza pandemic An outbreak of influenza that rapidly spreads around the world. See also *pandemic.*

isolation The separation of people known to have a contagious disease from those who are healthy. See also *quarantine.*

job action sheet The written responsibilities of healthcare professionals participating in an Incident Command System. See also *Incident Command System.*

just-in-time training The specific training for tasks that volunteers will be performing at the scene of an emergency.

Laboratory Response Network (LRN) The network of public and military laboratories organized into a three-level system for processing chemical or biological specimens in an emergency. The LRN responds to emergencies that involve natural disasters and to events that involve biological and chemical terrorism. See also *national laboratory, reference laboratory,* and *sentinel laboratory.*

litter bearers The people who move victims on stretchers.

Local Emergency Management Agency or Local Emergency Management Authority (LEMA) The community governmental agency that is responsible for public safety, emergency medical services, and emergency management.

man-made disaster A disaster caused by people or technology. See also *disaster.*

mass casualty incident An accident or event which results in enough fatalities or serious injuries to overwhelm the available resources.

mass prophylaxis The dispensing of vaccines, medicines, or antidotes on a large scale. See also *Point of Distribution.*

Medical Reserve Corps (MRC) A federal group that organizes and deploys healthcare volunteers. The MRC is organized by local communities to enhance the emergency and public health resources already in place.

medical waste The infectious or biological products that have no further use. See also *autoclaving, incineration, sharps waste,* and *steam sterilization.*

memorandum of understanding (MOU) An agreement among medical facilities to receive patients from one another in case of evacuation.

mitigation A phase in comprehensive emergency management that seeks to prevent or minimize emergencies. See also *comprehensive emergency management.*

mobile pharmacy A small pharmacy outside of the main pharmacy set up by pharmacists who bring with them a supply of the most commonly used medications.

mutual aid The agreements arranged with neighboring communities or institutions to assist each other in an emergency.

National Disaster Medical System (NDMS) The federal system of rapid response of medical teams, equipment, and supplies sent to a disaster area; moves ill and injured patients from the disaster to a safe area, and provides care for patients in hospitals away from the disaster site.

National Incident Management System (NIMS) A comprehensive national approach to managing all types of emergencies. Provides a national template that enables all levels of government, the private sector, and nongovernmental organizations to work together to manage emergencies.

national laboratory The highest level of the three-level system in the Laboratory Response Network. The national laboratory is responsible for isolating and identifying specialized strains of biological and chemical agents. See also *Laboratory Response Network, reference laboratory,* and *sentinel laboratory.*

National Laboratory Training Network A resource that provides laboratory training courses in clinical, environmental, and public health laboratory topics.

National Pharmacist Response Team One of ten teams that exist across the United States to quickly distribute and administer anti-infectives, vaccines, or medications in case of a terrorist attack or other disaster.

National Pharmacy Response Team (NPRT) A group made up of pharmacists and pharmacy technicians that is part of the National Disaster Medical System. This team can provide mass vaccinations or pharmaceutical treatments for large numbers of people.

National Response Framework (NRF) A guideline for how governmental, non-governmental, private sector, and local agencies will work together in responding to an emergency. The NRF ensures that the federal government can quickly deliver federal support to communities.

natural disaster A disaster caused by meteorological or geological events such as blizzards, earthquakes, floods, heat waves, hurricanes, and tornadoes. See also *disaster.*

Occupational Safety and Health Administration (OSHA) The federal agency responsible for workplace safety.

pandemic An infection that rapidly spreads around the world, causing large numbers of illnesses and deaths.

paramedic An emergency medical technician who has additional training and responsibilities, including emergency advanced life support treatment. See also *emergency medical technician* and *advanced life support.*

patient surge A sudden increase in the number of patients.

personal protective equipment (PPE) The protective clothing and equipment that reduces the exposure of a health care professional to a hazard.

Point of Distribution (POD) A temporary station for dispensing vaccines, medicines, or antidotes on a large scale. A POD is set up by a local government agency and is supported by federal resources. See also *mass prophylaxis.*

preparedness A phase in comprehensive emergency management that determines how a community or an organization will respond to an emergency. Preparedness includes organizing personnel, equipment, and supplies; training of personnel; conducting drills; developing communications and warning systems; creating evacuation plans; stockpiling supplies; and forming alliances with neighboring communities or institutions for mutual assistance. See also *comprehensive emergency management.*

Presidential Declaration When the Governor of a state requests that the President declares a federal disaster enabling the deployment of federal resources.

presumptive A positive test result for a suspicious biological or chemical agent.

push pack A ready-to-deploy container with enough medical supplies and medications to treat thousands of patients. See also *Strategic National Stockpile.*

quarantine The separation of people who have been exposed to a contagious disease and may be infected, but are not yet ill, from healthy people. See also *isolation.*

recovery A phase in comprehensive emergency management that includes activities and programs to help a community return to normal. See also *comprehensive emergency management.*

reference laboratory The second level of the three-level system in the Laboratory Response Network. Reference labs have the reagents and technology to conduct more sophisticated tests for biological and chemical agents than do sentinel laboratories. See also *Laboratory Response Network, national laboratory,* and *sentinel laboratory.*

respirator A device designed to protect the wearer from inhaling very small particles, such as viral particles or other airborne infectious agents, harmful dusts, fumes, vapors, or gases. See also *fit-tested.*

response A phase in comprehensive emergency management in which plans are put into action. See also *comprehensive emergency management.*

risk communication The communication of information about a health crisis to the public in a timely and accurate way that does not unduly heighten concern and fear.

scalable The ability to change the size of the response from small to large.

scope of practice The legal and professional definition of the training and responsibilities of allied health professionals.

sentinel laboratory The lowest level of the three-level system in the Laboratory Response Network. Sentinel laboratories are the hospital, clinic, or private labs where most laboratory technicians work and where most medical testing is conducted. See also *Laboratory Response Network, national laboratory,* and *reference laboratory.*

sharps waste The medical waste that includes hypodermic needles, syringes or tubing, scalpels, and pipettes. See also *medical waste.*

shelf life The amount of time a medication is usable before it reaches its expiration date.

sheltering-in-place To remain in a facility or other designated, protective area, during a disaster.

Simple Triage and Rapid Treatment (START) The type of triage that makes it possible for only a few rescuers, including those with limited training, to rapidly evaluate a large number of patients and move them to treatment centers for more detailed assessment. See also *triage.*

special needs patients The patients who need assistance in the management of their illness or condition and cannot be properly cared for in the evacuation shelters set up for the general population.

special needs shelters Evacuation shelters outfitted with special equipment and medical staff to care for people who cannot be properly cared for in the evacuation shelters set up for the general population.

staging area An area in which patients wait to be transported during an evacuation or other emergency.

standard and universal precautions The procedures established by the Centers for Disease Control and Prevention (CDC) to prevent the transmission of infectious diseases to healthcare workers and patients. Includes hygiene such as hand washing; protective clothing and gear; and procedures for handling sharp instruments and needles.

steam sterilization A method of medical waste disposal that exposes the waste to steam at sufficient temperature under a sufficient amount of pressure and for sufficient time to ensure that microorganisms are destroyed. Also called *autoclaving.* See also *medical waste.*

Strategic National Stockpile (SNS) A large stockpile of medicines, vaccines, antidotes, and related medical supplies maintained by the federal government.

terrorism The unlawful use of, or threatened use of, force or violence against individuals or property to coerce or intimidate governments or societies, often to achieve political, religious, or ideological objectives.

triage A procedure for sorting victims at a disaster site according to their condition and the resources that are available.

unified command An approach used in an Incident Command System in which multiple agencies work together to establish a single, coordinated plan. See also *Incident Command System.*

vendor-managed inventory The specific medications sent if needed for a surge in patients by the manufacturers who are under contract.

victims The people who are injured or become sick as a result of an emergency or disaster.

Index

Photo Credits

Chapter 1

p. 2, © Benjamin Lowy/Corbis; p. 5, Courtesy of Win Henderson / FEMA photo; p. 7, © Reuters/CORBIS; p. 9, © Brooks Kraft/CORBIS; p. 10, Courtesy of Butch Kinerney/FEMA; p. 11, Courtesy of Greg Henshall / FEMA; p. 12, Courtesy of Michael Rieger/FEMA; p. 13, Library of Congress Prints and Photographs Division Washington, D.C. 20540 USA; p. 14, © Jacques Langevin/CORBIS SYGMA; p. 15, © Orestis Panagiotou/epa/Corbis; p. 16, © Craig Lassig/epa/Corbis; p. 17, © Igor Kostin/ Sygma /Corbis; p. 21, Courtesy of Andrea Booher/FEMA; p. 26, Courtesy of Jocelyn Augustino/FEMA; p. 29, Courtesy of Michael Rieger/FEMA

Chapter 2

p. 40, Courtesy of Andrea Booher/ FEMA News Photo; p. 43, Courtesy of Andrea Booher/ FEMA News Photo; p. 47, Courtesy of Greg Henshall / FEMA; p. 52, Courtesy of Jocelyn Augustino/FEMA News Photo

Chapter 3

p. 64, Courtesy of Mark Wolfe/FEMA; p. 67, Courtesy of Win Henderson / FEMA; p. 70, Courtesy of Ed Edahl/FEMA; p. 73, Courtesy of Mark Wolfe/FEMA

Chapter 4

p. 96, Courtesy of George Armstrong/FEMA; p. 97, Courtesy of Win Henderson/FEMA; p. 103, © ELLINGVAG ORJAN/CORBIS SYGMA; p. 105, Courtesy of Andrea Booher/ FEMA; p. 108, © AP IMAGES; p. 111, James Gathany/ Centers for Disease Control; p. 113, Courtesy of James Gathany/ Centers for Disease Control; p. 117, © AP IMAGES; p. 121, Courtesy of Mark Wolfe/FEMA; p. 123, Courtesy of Manuel Broussard/FEMA

Chapter 5

p. 136, PRNewsFoto/Eli Lilly and Company; p. 139, Courtesy of Win Henderson/FEMA; p. 142, Courtesy of FEMA News Photo

Chapter 6

p. 148, Courtesy of Jocelyn Augustino/FEMA; p. 151, © AP IMAGES; p. 153, left, © Artiga Photo/Corbis; p. 153, right, © Luca DiCecco / Alamy; p. 157, © AP IMAGES; p. 160, Courtesy of Jocelyn Augustino/ FEMA News Photo; p. 161, Courtesy of Andrea Booher/ FEMA